greenhorns

"If we are very lucky, the Greenhorns will have their way and give America back its family farms."
> — Molly O'Neill, author of *One Big Table: A Portrait of American Cooking*

"Amid the many sad trends of our day, nothing makes me happier than the recent USDA proclamation that for the first time in 150 years the number of farms in America is on the rise. If you want to understand why, and if you want to feel the vigor and love now returning to the land, this book is the place to start!"
> — Bill McKibben, author of *Deep Economy*

"Right now, there is a whole generation of young, energetic, focused new farmers learning the craft, finding land, starting businesses. These are hard-working, dedicated, entrepreneurial folk, and all across the country they are producing exceptional food for their local communities. If you want to know what the voice of this powerful movement sounds like, this is the book for you. It is one of the most optimistic stories in America. The Greenhorns have put together a collection as satisfying as the food these farmers are producing."
> — Kristin Kimball, author of *The Dirty Life*

"The Greenhorns are my heroes. They are the face of America's new agrarians — mostly young, educated, ready for hard work, passionate, and rooted in community. I really believe they are our future, and we're damned lucky that they are. Their essays are insightful, thoughtful, and hard to put down. What a wonderful collection! I will recommend this book to everyone."
> — Deborah Madison, author of *Local Flavors* and
> *Vegetarian Cooking for Everyone*

"Hooray for the Greenhorns! Networking young farmers from coast to coast, they have become a formidable force in our new agricultural revival, spreading resources, sharing skills, documenting their experiences, and advocating for change in our agriculture system. These essays offer serious food for thought, especially for anyone considering taking up the hoe — no matter whether young or old."
> — Sandor Ellix Katz, author of *Wild Fermentation*, *The Revolution Will Not Be
> Microwaved*, and *The Art of Fermentation*

"This collection of young farmers' stories is a cornucopia of gifts from the earth. I learned with pleasure about a group of people who are turning the world into a much better place."
> — Les Blank, filmmaker

The mission of Storey Publishing is to serve our customers by publishing practical information that encourages personal independence in harmony with the environment.

Edited by Carleen Madigan
Art direction and book design by
    Carolyn Eckert
Illustrations by © Lucy Engelman

The information in this book is true and complete to the best of our knowledge. All recommendations are made without guarantee on the part of the author or Storey Publishing. The author and publisher disclaim any liability in connection with the use of this information.

Storey books are available for special premium and promotional uses and for customized editions. For further information, please call 1-800-793-9396.

Storey Publishing
210 MASS MoCA Way
North Adams, MA 01247
www.storey.com

Printed in the United States by
    Versa Press
10 9 8 7 6 5 4 3 2 1

Library of Congress Cataloging-in-Publication Data

Greenhorns / edited by Zoë Ida Bradbury, Severine von Tscharner Fleming, and Paula Manalo.
    p. cm.
 Includes bibliographical references and index.
  ISBN 978-1-60342-772-2 (pbk. : alk. paper)
  1. Farm life—United States—Anecdotes. 2. Agriculture—United States—Anecdotes.
  I. Bradbury, Zoë Ida. II. Von Tscharner Fleming, Severine. III. Manalo, Paula.
S521.5.A2G75 2012
635.0973—dc23

                    2011051495

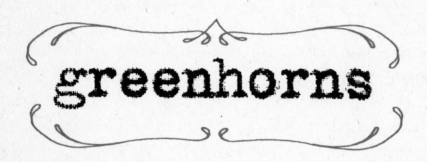

greenhorns

50 Dispatches
from the
New Farmers' Movement

edited by
Zoë Ida Bradbury, Severine von Tscharner Fleming,
and Paula Manalo

illustrations by
Lucy Engelman

contents

INTRODUCTION
SHOVEL SHARPENED, SHOVEL READY 8

contents CONTINUED

CHAPTER EIGHT

OLD NEIGHBORS, NEW COMMUNITY 217

RESOURCES 251

INTRODUCTION

Shovel Sharpened, Shovel Ready

The essays in this collection were written and selected by beginning farmers for the benefit of other beginning farmers, eaters, and aspiring agrarians of all stripes. These stories are narratives of production that we believe to be representative of the beginning-farmer experience in contemporary American life. Directly personal, they have given us the opportunity to reflect on the shape of the movement and the implications of our collective work, which has engaged us for varying amounts of time, in every region of the country. The authors/farmers represented here dream big in very different ways.

Americans, by and large, no longer think of themselves as part of an agrarian nation. That is, most Americans are neither engaged in agriculture, nor acculturated into its rituals, nor personally connected to its success. Trade, technology, retail — these spin the economic turbines of our major cities; these dominate the headlines of our newspapers, the avenues of academia, and the imagination of our televised pantheon. Though we are fed collectively, we have, as a culture, only just begun to consider the implications of our industrialized, and continually consolidating, food system.

Viewed through the lens of public health and climate change, in particular, this examination of what and how we eat has gained national prominence. Finally, people are paying attention to the food: where it is grown, how it is processed, who owns it, what degrades and pollutes it, the impact it has on our health. This conversation is long overdue; a third of our kindergartners are predicted to contract type 2 diabetes in their lifetime, and a third of climate change is attributed to the production, manufacture, and distribution of food. We've reached a crisis point. The consensus is almost deafening: The food system must change.

And who will make that change? Today's accepted rural narrative is one of crisis, abandonment, and attrition. For the past 30 years, as farms got bigger and prices spiraled ever downward, young people have been leaving agriculture and rural areas, and the rural culture has suffered tremendously for that loss. Small towns across the country stand empty and forlorn. There isn't enough money pumping through local businesses. There are fewer and fewer parishioners and contra dancers. Mega dairies, feedlots, processing factories, and grain elevators stand tall over an agricultural landscape that is ever less enticing and less accessible to ambitious youth. Country western to country eastern, the average American farmer is 57 years old and, likelier than not, he's on the brink of retirement or bankruptcy.

Thankfully, it is not just a cultural conversation about food that has begun in this country. There is also a quietly surging and committed movement of people who have recently, almost inexplicably, hitched their lives to agriculture. And just in time. Despite the paving of hundreds of thousands of pastures with either concrete or corn, there is still a thriving, driving need in the hearts and minds of a new generation of farmers to be makers of food, tenders of land, and protagonists of place. Bottom line: We want to and love to farm. We love to farm with such fervor that we are willing to jump high hurdles and work long hours to build a solid business around our love. It is not easy, or simple, or socially acceptable. Increasingly, though, entrants into American agriculture are propelled by a force of intention that ignores such things. Thirty years of protracted rural crisis have left quite a toxic wake: degraded soil, desperate economies, worn-down communities. With a mind for business and a brave agenda of personal, societal, and ecological sustainability, we've got a lot of reclaiming to do.

The work is difficult but it's relevant. Being or becoming a farmer is a thrilling act of creation; to do so, we hold a space between the present and the future, between ecology and humanity. We are directly involved in the reconstitution of a local, resilient, and delicious food system. This impulse to farm is felt by a wide (and widening) range of Americans. We are unified in our willingness to work, to change, to sacrifice comfort and social acceptance in the pursuit of an agricultural life.

The stories here are about that work: the intentions behind it, the structural challenges, the improvised solutions, the aches and the tedium, the bliss and the sheer physical toil. We all work very hard at our businesses — growing food, educating our communities, refurbishing barns, retooling equipment — and our movement benefits from the charismatic presence of hard workers; it sets a good example.

What is it to be a new farmer in America? Some of us are punky city kids, some come from suburbia, some come from families that taught handy skills, many are the offspring of recent immigrants. This is America, and it is colorful. We're proud of our diversity, our multicultural, multimodal, multipronged professional trajectories in agriculture. Some of us traveled east seeking roots, some traveled west seeking land. We hunted for old agronomy books in school libraries and used bookstores; we apprenticed ourselves to old ranchers and old hippies. We invent new ways of doing business in spaces designed by another era. We are Americans deeply in touch with our land, our place, and our role in re-forming that place. Our stories come from experience, but they vibrate with the anticipation of a particularly good American future.

Farming is an expression of patriotism and hope. Though our votes might be ignored in this country, as farmers we can still take pride in a nation we've directly cultivated. We can be proud Americans to the extent that we transform this country into a place worthy of that pride. We are still a nation of great lands and great towns. No matter where we were raised, we have come to the realization that it is our job to make it a remarkable future. And across all landscapes, suburbs, vacant lots, lease agreements, and lonely roads, we are crafting that future according to our own tastes. We stick our forks, tines, spades, and fingers into that particular part of the planet over which we have gained some governance. And, as a result, we eat well, we sleep well, and we earn the respect of our neighbors and families. Imagine: We can reshape the American landscape.

We do imagine, collectively, what our work adds up to: how it expands and extrapolates in the bellies and brains of those we feed and host and inspire on our farms. Our hopefulness has grandeur in it, not just sacrifice. We are creating our own new rural institutions of celebration and cultural interest: crop mobs and hoe downs and seed swaps. We reconfigure urban and postindustrial spaces with gardens in pots, start chicks in truck beds and swimming pools, erect temporary fencing, jerryrig air-conditioning units to keep our flowers cool. We recognize the power in DIY, the dignity of a small business, and the bold political gesture of it all. We can imagine already the outcome and consequences of our work together, the building and rebuilding of a sustainable agriculture, economy, and society. Ultimately, it's in our hands to make it so.

We continue to exert ourselves agriculturally. Difficult choices and professional acrobatics characterize that journey. It is hard. At each step, we must often radically reshape the economy in which we farm, and radically challenge the agronomic status quo. Perhaps you will agree that the insights of beginning farmers carry crossover value for other actors in our changing society, who must face down similarly daunting personal transitions.

Thomas Edison famously said, "Opportunity is missed by many because it is dressed in overalls and looks like work." Yes indeed, it is work. But it is work that must be done, and an opportunity not to be missed.

– Severine von Tscharner Fleming

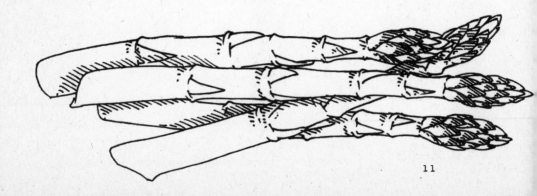

CHAPTER ONE

BODY—HEART—SOUL

Farming. It's physically demanding, requires entrepreneurial gumption, and tests our emotions.

When I first considered apprenticing on a farm, I was attracted to the agrarian lifestyle for its therapeutic and strengthening characteristics. Fresh, organic food would always be nearby, and I would be physically active — something that was lacking at my previous job. Still, I had some insecurity about diving into something so unfamiliar and uncomfortable. During my first farm internship, I was awkward with a hoe and tired quickly. Since then, I get stronger each season and have learned diligence, courage, and resilience.

Those lessons were reinforced this last winter during our third cool-season-vegetable CSA. Last fall and winter in northern California, rain hammered the soil for months with no mercy. Without a lot of greenhouses to protect our crops, we struggled to fill CSA shares. When it hailed in March, my heart sank. No matter what we did with the resources available, the weather kept beating us down. As tiny ice cubes pelted me and the crops, I took a long, deep breath and cried a little.

Thankfully, the CSA membership understood the challenges and agreed to postpone deliveries. We quickly built small high tunnels on the cheap, and by late April we started to have big beautiful vegetables again. If it weren't for encouraging words from CSA members, I probably would have given up on growing vegetables.

I love farming because I can feel its vibrancy. The farm is like a symphony of crops, livestock, soil, weather, equipment, and CSA members, and I, the farmer, am playing with them, being moved by the rhythm and enjoying the harmonies. With the challenges and lucky breaks that come along, I've cultivated persistence and capitalized on opportunities.

The intention of offering food with flavor and nutrition is motivation to farm and a reward in and of itself. In nourishing people around me, I see individuals come together around food I've grown, and I realize my place in the community. Laboring and producing within the finicky parameters of the earth and sky, I can share that sacred nature connection with others, and they support my partner and me as we grow our farm organism.

Delving into agriculture from an urban or suburban lifestyle is a big transition. The life of a farmer creates unexpected dynamics — both challenging and pleasant — among family and friends. On a small budget and with little time, taking care of our health and sanity is not always a priority, so customized self-care becomes part of the agricultural life. Every morning I stretch, and most days I eat the freshest produce from my fields and enjoy delicious meat from animals I've raised.

How do we stay steady on our epic adventure through the seasons of surprise, frustration, elation, exhaustion, hilarity, defeat, and miracles? The following essays provide a glimpse into individual journeys and address the importance of our bodies, hearts, and souls — all dedicated to farming. As we develop as farmers, full of hope, we stick with it and grow, making our bodies stronger, staying true to our hearts, and with our souls evolving.

— Paula Manalo

The Physicality of Farming

BY JEFF FISHER

Born and raised in Cedar Falls, Iowa, Jeff Fisher became interested in sustainable agriculture while attending graduate school in Olympia, Washington. Jeff is in his second season of farming at Cure Organic Farm in Boulder.

"Your hands are going to bleed."

Anne Cure, owner of Cure Organic Farm in Boulder, Colorado, said this softly while looking off into the distance as Jack, one of the other farmers, described the day's task of transplanting thousands of seedlings from the greenhouse into the field. The "bleeding hands" comment was not ill-natured in any way; it was merely a statement of fact, one learned through many springs of transplanting thousands of seedlings into the field. This was the acknowledgment that today the fields were going to be especially tough to plant. It would be a painful process for a new farmer's hands.

Cut, cracked, and bleeding fingers are just the start of the physical hardships of farming. Over the course of my first full season, I spent long days bending, squatting, grabbing, twisting, pulling, pushing, cutting, walking, running, jumping, dragging, digging, pounding, lifting, tossing, catching, and reaching. At the end of each day I was left with aches, pains, cuts, cracks, blisters, infections, stings, and sprains. Early in that season, I always seemed to have nicked fingertips, because of my inexperience at cutting greens with a steak knife. Cracked fingers were a mainstay during the cold days of spring and fall. Our honey harvest in August brought a very swollen right foot and ankle from a host of bee stings. A marvelously infected big toe in September grounded me for three days. In October came the grand finale of farm injuries: I badly sprained my ankle after leaping over a bed of arugula and landing awkwardly. This was followed by a month of physical therapy.

My first year of farming provides these examples of the very physical toll that farming takes on one's body, but there are also the constant sore hands, feet, and back that come and go.

I'd never thought about being injured, even as I was gracefully soaring over that bed of arugula. I definitely started thinking about it when I landed and felt my ankle twist inward under the weight of my body. I knew exactly what happened the second I landed; I knew this was going to be painful for days and maybe weeks. I also figured I still had a couple of hours before walking became really difficult.

Six hours later, after completing the day's restaurant deliveries, I was barely walking. Besides the pain in my ankle, the reality of my injury began to set in. What if the ankle was broken? How long would I be unable to farm? Would I still get paid while I was injured? Had my first full season of farming just come to an inglorious end?

I didn't have a lot of options for logical next steps, other than to allow myself time to heal. Later, taking a proactive approach and preparing my body for the season would help prevent farm mishaps. I could also avoid injuries by staying focused on the task at hand, wearing good shoes, and, of course, taking care when jumping over beds of greens.

The simplest preventive measures were to start stretching and to change positions frequently. I'm still amazed at the ways my body must contort in order to plant seedlings and harvest vegetables. One moment I'm in a deep squat, then I'm bending at the waist. Next I pull out the wide stance, followed by the one-knee-down position. Variety is the name of the game with many farm tasks. Even the most adept farm yogis (which I am not) can't hold the same posture for more than ten minutes, so I found myself constantly changing it up.

I also quickly learned the importance of adequate sleep and eating well. A poor night's sleep or a missed breakfast leaves me less alert and less effective during the day; this is when accidents occur and injuries happen. Although aches and pains throughout the season are often inevitable, it's important to take care of your body, as you would any farm implement.

After spending a season harvesting tens of thousands of pounds of vegetables, I can attest to the physical investment in the food we eat.

I'd never thought about being
injured, even as I was gracefully
soaring over that bed of arugula.

To farm is to engage fully in a truly physical way of life. In the course of farming, we use plenty of trucks, tractors, and tools to accomplish our tasks. Where I farm, though, our scale of production is small enough that we will never acquire the equipment to automatically cut greens, dig carrots, or pick tomatoes. My body is the implement that's going to plant, weed, harvest, wash, and pack every last pound of produce. At times my body seems to be rebelling, practically screaming at me to stop and leading me to question my pursuit of farming. Will my body be able to do this sort of work for twenty more years?

So far, any question about whether or not farming is my way forward has been answered with a resounding yes. I want to help change the practice of agriculture, feed my community, and grow beautiful, delicious food. The cost might be a few bumps and bruises, but for me, the physical investment is worth making. ✀

Farmer-Mama

BY SARAH SMITH

With her husband, Garin, Sarah Smith owns a certified-organic operation in
Skowhegan, Maine. They milk forty cows, selling to the CROPP Cooperative, and offer
organic beef, pastured broiler chickens, eggs, and seasonal CSA shares. Sarah
also manages two acres of mixed vegetables while raising the couple's two children.

It's a sweltering night in late July. Just as on so
many other nights, I lie awake, one thought rolling into the next.
Even with the window open wide, the air is still and stifling in this old
Cape farmhouse. I roll onto my stomach, attempting sleep one last
time. As I begin to nod off, I hear the faint cry from the room adjacent: "Mama? Mama?" Each call is progressively louder until I finally
stumble out of bed and across the hall. It's Cedar, my four-year-old,
calling me.

This is a child who has not slept through the night since the day
she was born. I soothe her and stagger back to my bed twenty minutes
later. It's the second time tonight she's needed me to pull up the covers
or get her some water. Now, as I drift off, I hear a cry from the crib in
my room. I hesitate for a minute or two before I stand and cross the
carpet. My sixteen-month-old, Reed, is quiet once in my arms as I
carry him back to my side of the bed. The bad habit of bringing him
into bed is a symptom of exhaustion. I can't bear to stand there nursing
him in my state. Lying down together, his belly is filled and my mind
is too, and we're bathed in the light of the moon outside my window.

Dawn comes early, around five o'clock, and with it the blare of the
alarm clock. This time of year you can't hit the snooze button as you
might in November. The day ahead of us is full. My husband, Garin,
and I slip out of bed, not speaking a word but each dressing in the
early-morning light. I have to sneak out quietly so as not to stir Reed
in our bed. By five fifteen, Garin is headed out to bring in the cows for
the first milking of the day and I've put on the coffee, the dogs and cats

are fed, and the monitor is turned up to listen for the kids to awaken. The next two hours are my time. The children are asleep and I get to work, not bothered or harassed. Even the interns are still sleeping.

The stacks of coolers around me in the kitchen are filled with dirty glass milk bottles. I glance at the clock, knowing I must get on it. The kids will wake up and then my progress will shut down. I scrub bottles as quickly as I can. By six fifteen, our interns start to wander in for a visit to the bathroom and to prepare breakfast. They must be ready to go by seven, so this hour can get a bit tight with all of us sharing one bathroom and one stove. I continue to wash bottles as the toaster pops and eggs sizzle on the stove.

It's seven o'clock now, and the ladies are sitting around, sipping coffee and chitchatting. Jumping into the conversation, I always start with: "All right, after you're finished with the chickens . . ." They know this is how the day begins. I go on, telling them to head to "Boston" to harvest for market. The primary production garden at our farm earned that name when we told Cedar a couple of years ago that we were headed to the "far-out" garden. She said "We're going to Boston!" with so much enthusiasm that the name stuck. The women are going to harvest new potatoes, onions, salad mix, broccoli, kohlrabi, tomatoes, peppers, and so on. Off they go, list in hands, and I watch out the window from the sink, where I continue to scrub away at the smelly yellowed milk bottles. I watch as they load up crates and baskets for the harvest. A part of me longs to join them, but the kids cry for breakfast and there are a few bottles still left to scrub. It's often this way, the kids wanting or needing me for something and the job I've started not quite done.

Fed, washed, and dressed an hour later, I load Reed into the backpack with Cedar following; we're going to bottle milk. An hour passes and forty gallons are bottled. Loading all the coolers of clean, empty milk bottles is chore enough even without thirty extra pounds strapped to my back. Lifting a cooler filled with seven gallons of milk in glass bottles, with the thirty pounds in tow, is a good strength-building exercise. We finish putting the coolers into the van just as the interns are ready to load up the crates of clean and bunched vegetables for market.

It's ten thirty and the stress starts to hit, as I know we need to be out of here by eleven thirty at the latest. There are lunches to put together, diapers to change, water and sunscreen to pack. Now add to that filling the beef cooler, grabbing the extra milk out of the fridge, getting the eggs washed and boxed, and packing the market box. In my mind, I'm constantly reviewing the lists of all the things I can't forget. I'm always running around on Tuesday mornings, tired and rushed, driven by caffeine and the ticking of the clock. It's hard for the kids to be hustled around ("Get dressed, get your shoes . . ."). I sometimes bark commands at them as I would at the dogs. I rely on my good interns to aid me in tying up all the loose ends as we load up: buckle the kids, grab market bags, and don't forget the pack-and-play.

Each day is like this for a
farmer-mama. There are no vacations,
Saturday gymnastics classes,
or afternoons at the playground.

The kids and I are five minutes late in hitting the road, which is no surprise. We live by the five-to-ten-minute-late schedule every day. I have a few deliveries before we get to the market, which is forty miles away. I'm prepared for the drive. At each stop, Cedar will ask to get out and I'll say, "No, we're late, and I'm just going to rush in." She and toddler Reed will fight about the water bottle and need to have food before I'm ready to serve it. I'll carry a weighted cooler into each store with a smile even as I can hear them screaming at each other in the car. I mentally deny that my life is more difficult than any other mother's, but the truth is that each day is like this for a farmer-mama. There are no vacations, Saturday gymnastics classes, or afternoons at the playground.

We hope someday to have a farm that is sustainable enough that we can afford the help to cover for us and enable our children to have a life outside of work. For now, we believe we're providing them with amazing experiences that many other children don't get — a relationship with livestock, lots of fresh air and sunshine, terrific math skills (making change at the market, for example), great social skills from being with people, and so much more. But on many days, all this comes at a cost to our family.

After deliveries, we have about a half-hour drive to the market site. There's no radio in this old Dodge van, so usually the kids get quiet and nap. I'm tired from the night of lost sleep but push on, as I do every other day.

Setup takes almost an hour this time of year, with so many beautiful vegetables to display. I pray each Tuesday as I pull into the parking lot that the children will sleep until I'm ready to sell, but the quiet click of the door always stirs Cedar. Sometimes I get lucky and can sneak her out without disturbing Reed, but not always. Today my prayers pay off, and I get her out in silence. The market is her playground, her territory, so she immediately takes off to get an apple from the orchardist or to help the manager set up signs. I start pulling out tables, coolers, and crate after crate.

Of course, halfway through Reed begins fussing. Time is running out. Customers always start to arrive ten minutes early and I'm never ready. I pause in my setup to get out the pack-and-play, which I stick Reed into with a few toys, hoping this will work at least temporarily.

Potatoes, tomatoes, greens, carrots, beets, and peppers need to find their space on the table. I'm a decorator of sorts as I listen to Reed fuss and watch out of the corner of my eye to make sure Cedar hasn't run off or bothered people intent on setting up their own booths. Just as I predicted, at ten minutes before two o'clock some customers wander in. I hate to sell before I'm ready, but I do it week after week. I haven't yet put out the price signs or topped off the basket of summer squash, but I greet them with a smile.

"Hey there, what can I get for you today?" I say, as I scoop Reed over my shoulder and back into the backpack. They say, "Oh, take your time," or "I can come back," but I shrug it off and tell them I'm fine and ready to get them what they need. As the afternoon rolls on, I juggle the selling, letting Reed run around, conversing, topping off baskets as things get low, and putting empty crates into the van in hopes of getting a head start on the evening pickup. Cedar holds her own even if on occasion I have to remind her not to get in someone's way or to give her brother some space. Sometimes she's a big help by pushing Reed around in his stroller. Most of my customers are very understanding when I have to put them on hold to deal with a child. I appreciate that they know they're supporting my family in so many ways.

It's six in the evening now, and the market is over. Pickup is faster than setup, but it's still slowed by the kids, who are now turning into pumpkins. They're tired and hungry, but this is what we do, and like it or not, they have to deal with it. Halfway through pickup, I buckle them in to keep them contained and away from the moving trucks and vans. They sense that I'm trying to finish so we can go home, and I think this helps them to be more patient.

Not long after we leave, the van quiets as they drift off for a late nap. I sing them a song. It is nice to have peace for the next half hour.

By the end of the day, I feel fulfilled and worn out. It's rewarding to raise food for people, which they feed to their children to nourish them and help them grow. I love to spend time explaining why local, why organic, why my farm, and how to cook or process something unique. I've changed people's choices and, frankly, their lives, and that's empowering. My children help me tell that story at every market. Whether it's by Cedar selling onions one day or Reed's glowing chubby cheeks and eyes that are the epitome of health, our story is beautiful and amazing.

Even in our most challenging moments, as we go about our daily business, I know that every mom has these moments. I'm different, though. I'm a farmer and a mama; both are full-time jobs and both are the most difficult jobs in the world.

Coming home, I find that my interns have prepared dinner while I was away, and they help me unload the kids and the leftover food from the van. At eight o'clock (sometimes later), I enjoy a glass of red wine with dinner and start to relax. Bedtime has challenges, too, and I know that I will be awakened several times throughout the night, but I still feel good. I'm exhausted as my head hits the pillow, but I have a belly full of the world's best food, a hardworking husband beside me, two amazing children, three hundred beautiful acres to steward, animals to care for, and people to feed. ✑

Doing. Instead of Not Doing

BY EVAN DRISCOLL

A father, farmer, and writer in Portland, Oregon, Evan Driscoll owns Sasquatch Acre, where he grows more than a hundred varieties of organic vegetables and raises thirty laying hens.

In one night they vanish: thousands of seedlings consumed a month before our first farmers' market. Panic takes hold. I begin to sweat hard. Instinct carries me down the street to the feed and seed shop.

Waiting in the checkout line, I run into my farm neighbor, Nate.

"I think it's slugs," I suggest. "It's weird, though, because they dug down and ate the seeds, too. How do they do that?"

"Well, they don't," he informs me. "Sounds like you have voles. Or mice."

"Huh."

I buy the fifty-pound sack of salt in my arms anyway.

Quitting an awesome desk job was a difficult thing to do. My hours were predictable, the work was predictable, and my mind was at relative ease. Post-work hours and weekends were dedicated to my infant child, times we spent at parks, cafés, and just wandering through neighborhoods.

But circumstances changed, and off we went — my girlfriend, child, and I — from Austin, Texas, to its northwestern equivalent, Portland, Oregon. My girlfriend was enrolled in law school there, and our new location held no prospects for me, few friends, and no connections. A trying time lay ahead.

I watched a depressed economy swallow hundreds of my job applications. I was left with little to do but childcare and thumb twiddling. Lots of thinking.

Thinking.

And such.

And nothing.

And then, eventually, something.

I f**king hate farming. It's so stupid and I hate it, and it's really hard and I don't know what I'm doing and I'm not making any money (maybe breaking even) and there are all these *things* everywhere . . . these f**king things . . . why do I need all these things to farm? And all these people wanting everything, and *I* want everything, writing these words that try to mean something but mean nothing. I'm a failure. I'm twenty-five and directionless. Farming? *What the f**k* was I thinking? Back to grad school — no. F**k school.

Standing, I lean forward on my garden hoe, hard. It stabs me just under my shoulder in a satisfying way.

There's a blue heron, gliding.

I have no idea what's going on right now.

I do know I'm pulling weeds.

The decision was made: time to farm. It just made sense. This, I thought, would be the most direct way to have a positive impact on the environment. I sure wasn't going to help develop cheaper and more efficient solar panels with my liberal arts degree. I don't have a track record of being a charismatic activist, as I have considerable difficulty organizing even a potluck with friends. Then, farming. Environmental stewardship at its most basic.

I quickly found that, without experience, I would not be making any money working as a farmhand. I must have called every small farm within a twenty-mile radius of Portland. A typical conversation:

"Hello, this is _____ Farm."

"Hi, there. I'm looking for farming opportunities, and I was wondering if you have any paid apprenticeships. See, I need to have some sort of income 'cause I have a kid and . . . "

"Nope."

"Thanks so much!"

Luckily, more *unpaid* opportunities were available than I could shake a stick at. I found one that would fit my life: twenty hours a week of farmwork, on top of my current forty hours at moneyed jobs, and childcare to boot. Reasonable, I thought.

And there I was a month later, dropping peas into cold soil. Over the course of the next six months, I would learn to build a compost pile, make a greenhouse burst with vigorous seedlings, build and run a CSA, cook quick and healthy meals, manage volunteers, schedule plantings, identify plant diseases and problems, and manage nutria.

"Nutria problem," my farm sensei tells me. "I know there's a tax to pay to nature, but these nutrias might be taking more than their fair share." They've eaten half a bed of lettuce mix, a considerable amount for a half-acre farm. My sensei called up some urban trappers, the kinds who use traditional, tried-and-true capturing methods.

"I think they eat what they trap," she ventured.

Now all the tomatoes are dying. Beautifully poetic. The most symbolically important crop on the farm refuses to grow properly. It's May, and they've been stagnant at two true leaves for almost a month now, and we are clueless as to what to do.

"There's not enough ventilation in here," I tell Travis, my cofarmer. "We need to build a window on the other side of here."

And we do, and the tomatoes go on stagnating.

These tomatoes sure are sassy. Maybe this is how they grow. Or maybe they're taking a break before they freaking explode with growth in summer. That's probably it. I'm confident.

Whatever. I hate farming.

And off we go to dump 432 tomato seedlings into the compost pile. That's roughly fifty dollars' worth of seed and a few bucks of potting soil — a large chunk of change to us.

"You throwing those out?" Victor asks. He's our landlord's farm manager.

"Yeah, they've got some sort of disease." I say, on the defensive.

Victor takes a tray from the stack I'm holding. His eyes narrow. He tilts the tray, providing another perspective. Tilts it carefully in another direction. Brushes the tops of the seedlings roughly with his palm. Takes a pinch of soil.

"What's the soil?"

"Equal parts compost, perlite, pumice, and coconut hull. Made it myself."

"This is sawdust."

"Oh."

"Spray these with fish emulsion twice a week and see what happens."

With fish emulsion, they blossom into massive, magnificent seedlings, begging to be transplanted.

Nitrogen deficiency. Huh.

I sat in front of a window that left no room for daydreaming — doing so would only bring about a dream that consists of both drizzle and gray. January in Portland is rugged. Not ice-covered-streets-freeze-to-death-blizzard-snow rugged, but drizzling and gray and constantly these two things at all times without exception. So I turned my eyes to the computer screen in front of me. I would be graduating from farm apprentice to farm owner in a short time. The first steps I took to prepare were the following non-farm things:

- Build a website
- Make business cards and a banner for market
- Research and make seed purchases
- Build restaurant relationships
- Apply to farmers' markets
- Gather lots of physical resources
- Make a farm plan
- Make a business plan
- Make a planting schedule
- Crop profiles
- Budget

Simply creating the farm plan put me in the mind-set that I was on the land. It helped me make as many decisions off the land as I could so that I wouldn't have to make them on the land, where communication is both timely and costly. When all was said and done, the plan ran more than a hundred and twenty pages.

It was possibly the most valuable/worthless document I've ever created: Once finished, we rarely, if ever, referred back to it. Reality and planning are two very different things.

And then, holy sh*t, we're selling our vegetables *to people*. And these people are *happy*, and we're happy, and they *love* our salad

mix— the edible flowers are a hit. I can't believe the sun is out — it
hasn't been out in, what, two months? And now look, there it is.
All the other vendors are just lookin' so good. What a fine market —
our first market, yes, but I know, I know we're going to find something
profoundly spiritual here.

I don't live on my acre. I have this big metal box with a huge tank
that holds something called "gasoline," and this propels the box to
a destination of my choice, within reason. I use it to get to my farm.
This commute, however, is not by choice. My four-year-old son
requires most of my attention and time, as my girlfriend — his mother
— is in law school. She's present as much as she can be, but most
of the time she can be found surrounded by crisp stacks of 8½-by-11
paper, highlighters, and little colored tabby things to mark special
pages in obese books.

It was time to farm. It just
made sense. I sure wasn't going to
help develop cheaper and more
efficient solar panels with my
liberal arts degree.

Food suddenly became a big thing in my life when I was forced,
unwittingly, to cook the majority of my family's meals. I guess I didn't
realize that having a girlfriend in law school would mean that I would
become the primary caretaker of our child, and, thus, a responsible
adult. Cooking, laundry, drop-offs, pickups, making sure he doesn't
wear poop-stained pants to school — these are worn only to his
uncle's house — ensuring that he consumes enough calories so as not
to perish, ensuring that he consumes enough water so as not to
perish, ensuring that he does not ingest chemicals, or bump his head

on a corner, or follow my example as I scream obscenities at ten beds of potatoes, tomatoes, cucumbers, and squash that are all under-performing/dying. This helps them grow.

In short, my life is a balance between farm and family. It's difficult to really dive in and make a go of it doing it this way. But at least I'm doing.

We're doing our damnedest to not work wet soil. As first-year farmers, we're doing everything by the book. We, in fact, know no other way of doing things. Don't work the soil when it's too wet, all the books repeatedly repeat. You will destroy your soil structure. You will destroy the diversity of the life within. You will destroy yourself and your loved ones.

And here we are, digging deep into sopping wet, heavy clay soil. It's mid-May and our first farmers' market is coming in a quick six weeks. We need to bring something. Coming to the first market shorthanded would be disastrous. We will be red-cheeked. Labeled amateurs SO FAST by customers. FIRST IMPRESSION RUINED. Mocked by vendors, spited by God himself, with his shining, mas-sive, iron fist, gleaming brightly in the sky above. We must bring something.

The bed-building process is incredibly laborious. We're digging into our sixteen-inch pathways a spade deep and flipping the soil onto our four-foot-wide beds. At seventy-five-feet long, and with the soil as wet as it is, our muscles are screaming at us to stop. What results is a blanket of gray mud bricks the length and width of our beds. At least the weeds beneath aren't going to break through.

We've now dug in ten beds this way and are waiting for a day or two of sun to slightly dry the mud cakes. Those days eventually come, and we borrow a neighbor's rototiller to break them up a bit. They now look like little pebbles, and we layer this with a good three to five inches of compost. We use a total of fifteen yards.

We drop some seeds directly into the compost, cover them, and pray for even the slightest amount of growth. And grow they do. And we bring head lettuce, lettuce mix, radishes, mustard mix, garlic scapes, and eggs to our first market. We sell out quickly, and folks are ultra-happy. One year later, we have yet to till these beds. There have been three successions of veggies in some, and counting. The soil is improving, and seems to be teeming with life — lots of worms and little buggers.

That said, I don't recommend working wet soil.

I do advocate wingin' it.

On his way out to the farm, Travis tells me he sees a bald eagle and two golden hawks having an epic three-way battle over a bridge. Three looming mountains comprise their background, he says.

My boy stands in the middle of our first successful broccoli patch. The air is silky warm. Gold sunbeams bounce off his soft cheeks.

I walk naked around the land in utter darkness. ✠

You Are Not Alone

BY MEG RUNYAN

Meg Runyan runs Wild Goose Farm, which is part of the Farm Business Development
Center at Prairie Crossing in Grayslake, Illinois. She sells her organic produce through
a CSA and a farmers' market. Meg farms 3 acres with an employee and a small crew
of volunteers, including her parents.

I walk to the greenhouse in early-May sunshine.
The day is just beginning and already my mind is filled with tasks
that need to be done. I'm anxious to start on my endless list. There's
just too much to do today. Before I open the door to the greenhouse,
I pause to feel the gentle breeze, listen to the birds, and taste the
cool spring air. A moment of pure joy before the work begins.

The warmth of the greenhouse greets me with the lovely scent of
growing things. Quickly, I set up my seeding station, gathering potting
mix, trays, and seeds. It's lettuce and squash today, according to my
schedule. I settle into the methodical chore of seeding. As I work, I
think of those who, at this moment, are driving to their various jobs.
My life must be so different from theirs.

Before I know it, my greenhouse time is up. I quickly water the
new trays along with the rest of the greenhouse. I close the door and
hurry to meet my fellow farmers for our monthly field walk. Today,
we're doing a field walk of all our farms, and will share our ongoing
plans and ask questions. Matt is already there, waiting for us. He and
his wife, Peg, run a well-established organic vegetable farm in Prairie
Crossing. The rest of us are "incubator" farmers. We rent land, greenhouse and cooler space, and tilling equipment from the Farm Business
Development Program at Prairie Crossing. Matt and Peg serve as our
informal mentors.

I join Matt at the washing station and watch the other people
arrive from various parts of the farm. Jeff, a fifth-year farmer, drives

his tractor in from doing some early-morning cultivating. Alex, a second-year farmer like me, walks over from another greenhouse. Eric, who manages a school farm, yells out to us from his group of students that he'll catch up with us later. Nick, a first-year farmer, comes in to start his day with the field walk.

We start the tour with Matt's fields. We pass rows of healthy plants. Off in the distance, we see his crew transplanting. Matt shows us his beets, a vigorous crop with few weeds. Then he turns to where his peas should be. He cringes as he shows us the empty beds. Hardly anything has germinated. Looking at the forlorn beds, I realize that even experienced farmers have crops that fail. I am both sad and comforted.

We move onto Alex's fields, where he's working on a movable hoop house. The ribs and base are up; he's just waiting for the plastic, so he can start growing tomatoes inside. We're full of questions, as we're all in different stages of constructing our own movable or semipermanent hoop houses.

Jeff takes us to his fields next, which he has just cultivated with his tractor. Jeff asks Matt about his tractors and cultivating tools. This is Jeff's third year with his tractor, and I think he's finally enjoying himself. He's struggled the past two years, spending more time adjusting and fixing the tractor than the time it was supposed to saved. It's been a tough learning curve, but he's come a long way.

Now it's Nick's farm. We visit his pens of turkeys and chickens. Nick is planning on selling meat and eggs, along with his vegetables, all in his first year. I'm curious to see how he fares with his turkeys: I'm interested in turkeys, too. Eventually, that is. I'm not ready for livestock just yet. The vegetables are demanding enough.

Finally, we come to my fields. I was anxious earlier about showing my peers the fields of crooked rows and somewhat weedy crops. But, touring the others' fields, I discovered something: They have crooked rows and weedy crops, too. There's no reason to be bashful.

I proudly lead them to my garlic, growing tall and green. I'm especially excited, because this is my first garlic crop. Planting them the previous fall felt like practicing faith. "All right, little ones," I said as I tucked the cloves into the soil. "I'm trusting that you'll come up in the spring." Now, in May, I almost giggle every time I look at my garlic.

> I laugh every time I stop with a hoe in my hand to text the other farmers to see if a tractor is free.

As we walk around the rest of my fields, we talk about tilling. I'm more comfortable on the tractor now, but I still have a lot to learn. I'm curious about how others are prepping their fields and dealing with compaction. We beginning farmers share a tractor, a few implements, and a walk-behind tiller. As a result, we've become skilled in texting. I laugh every time I stop with a hoe in my hand to text the other farmers to see if a tractor is free. They're always practical messages, though. I have yet to text anyone, "OMG! I h8 mosquitoes!"

While we're talking, Jeff and I discover we both want the tractor this afternoon. We quickly work out a plan: Jeff will get the tractor first, while I do some fieldwork. Once Jeff is finished, he'll drive over to my fields. He'll take my truck back to the shed and I'll put the tractor away when I'm done.

With the plan set, we all head back. I gather the flats of lettuce, kohlrabi, and scallion seedlings I need to plant today. Each tray is soaked with a smelly concoction of fish emulsion and water. I load the truck with the plants, row markers, and hoes for weeding. I take a moment to check off the tasks completed for the day, enjoying the satisfaction each tick brings.

The drive to my fields takes me past the others. I see Matt working with his field crew. Oh, I wish I had a crew helping me today! Maybe I wouldn't be so dead tired and achy at the end of the day. Next, I see Alex with Alison, his wife, who will spend the rest of the day working in the fields. Jeff has joined his crew of two, continuing the weeding and planting they did during our field walk. I wave as I pass Nick in his fields. He's alone today, too, but his girlfriend will come out later in the week.

I reach my fields. I grab the row markers and trays of lettuce. As I alternately mark the rows, pop out lettuce seedlings, and tuck them into the soil, my mind wanders to the others farming around me. My mind fixes on what they have that I don't.

They all have someone with whom they can share at least some of the burden of farming. Some even provide an outside income. I feel the envy grow inside me. Why couldn't I have found someone to share this adventure and this burden? I feel so alone sometimes. It's overwhelming to have every decision weigh on me. Figuring out how to run a business itself is daunting, let alone learning to farm and run a business. How did I get myself into this? Why did I choose farming? I just wish I had some company. Frustrated and full of self-pity, I finish the lettuce with a huff. I grab a tray of scallions and plant them next to the lettuce.

As I work, my tantrum begins to subside. I remember that my mom came up yesterday and helped with planting and weeding. I think of how supportive she and my dad are. They've sweated and labored with me, and have given me a loan. I guess I can't claim I'm alone in my farming endeavor. I realize I won't be alone tomorrow either. A new employee and a volunteer are coming. In fact, just last Sunday, people in my church were asking if I needed help and how they could support me. If I honestly think about it, I'm not alone at all.

My tantrum is over now. Perhaps it will return another day, when I feel overwhelmed with responsibilities. But for now, with the reminders of friends and family offering real help and support, I have a sense of peace. While I focus on these calming thoughts, I hear the roar of a tractor. I look up and see Jeff driving into my fields. I look at my truck and see the unplanted kohlrabi still there. "Oh well," I think to myself, "transplanting will be a good experience for the people who are coming to help tomorrow."

I brush dirt from my knees and gather the empty seed trays and the markers and head over to Jeff. We talk for half an hour about different crops, the weather, farmers' markets, and insects. Farmers can be a bit chatty. We make the switch, Jeff climbing into my truck as I climb onto the tractor. I watch Jeff drive off and think about the field walk this morning. I realize again that I'm not really on my own.

I never would've guessed that my life's road would lead to farming. My journey learning to farm has been just as unpredictable. When I started, I didn't know the Farm Business Development Program at Prairie Crossing Farm existed. I didn't know of the mentors and friends I'd meet along the way. I took a chance and followed this newborn passion for farming. Though I don't know where exactly I'll end up or what my farm will someday look like, I do know that opportunities will arise. ✍

Two Pigs and True Love

BY ANDREW FRENCH

After growing up in small farming communities in the Midwest and enjoying the many benefits of country living, Andrew French escaped to Minneapolis to become a chef and landscaper, only to learn that his real passions were farming and permaculture. He's living the dream with his fiancée, raising ducks, goats, pigs, vegetables, and mayhem.

It was a wet spring afternoon, and Khaiti and I were hunkered in the snug little red barn on her farm, watching Metallika, the sweet, earless purebred La Mancha goat, with parental concern.

"It's happening now!" Khaiti yelped. She got closer to the goat to help her if needed.

With groans of pain from deep within her throat, Metallika emitted a slimy creature into the fragrant hay. It was a beautiful buck that Khaiti promptly named Two Tone. I uttered a single word: "Wow." I had thought I'd videotape the whole experience, but this thought evaporated as the next gooey goat baby squeezed out of Metallika's hind end. It was a devastatingly cute female with dark coloring. Khaiti named her Trixie. We crouched there in an enthusiastic huddle near the afterbirth and goat poop, and chattered like our jaws were spring-loaded. I finally whipped out my camera and filmed a little snippet of the newly born and surprisingly energetic Trixie jabbing at her momma's swollen teats to get her first taste of milk, the colostrum. Two Tone stood there in a stiff-legged daze and seemed amazed at this momentous change, being on the outside of his momma for the first time. My heart wobbled as he took his first shaky steps.

The forecast was for a downpour that night. I knew driving my old truck in the dark in the rain was dangerous, so I decided I'd better get home soon. I didn't want to go, though. I felt like I was already home. I wanted to snuggle up with the newborn goats in the hay and go to sleep with the rain pounding on the metal roof. I was torn but decided to play it safe. I said my good-byes to Khaiti and the goats with a heavy heart. Sure enough, it started to pour. I was forced to drive at a snail's pace all the way back, giving me time to reflect on all that I had seen and felt that evening.

To make some sort of sense out of all the emotions coursing through me, I decided to do something artistic. I was going to take that little snippet of video and play a little song as the soundtrack. I picked the song "Unravel," by Bjork, because it was the most beautiful song I could think of to illustrate the intensity of the experience. I got out my electric guitar and hashed out a version of the song that I thought was satisfactory. It was late when I finished, and before plowing into bed for the evening I e-mailed Khaiti the little video. I thought she would really enjoy it.

A few days later, we were chatting online. There was a lull in the conversation, and then she asked me if she could tell me something crazy. She explained that she had a crush on me. Watching the video had been a pivotal point for her in how she felt toward me. I panicked and ended the chat abruptly. I went outside to smoke a cigarette and think things through. I was worried by this change in our relationship, and I didn't want to lose an amazing friendship to a romantic interlude.

I told her in an e-mail that I didn't think it was wise to pursue this; I didn't want to wreck the relationship we had. She responded by asking if she could come visit me. I parried with it might be too weird. She said she was going to come over anyway. She did come over, and we laughed at the awkwardness. We talked it over and decided to give it a try. And so we began to farm-date.

On a farm, things don't stay clean. "Getting dirty" takes on a different meaning when you're farm-dating. When you have animals, you get manure on everything and everywhere, and if you're not comfortable with that, then farm-dating is not for you. Farm-dating is when you lie in the grass with your beloved by the gardens and animal paddocks and let the smell of fresh earth and green things mingle with the stink of manure rotting in hay, and you consider that the best aphrodisiac on the planet. It's when, instead of dinner and a movie, you both go out and put up more fencing for the goats. Instead of sending her flowers, I turned over the soil in her garden with a digging fork and took out some saplings.

Meanwhile, back on the home front, my small landscaping business was starting to take off. I had wrangled up some good clients and projects, and I felt that I was going to reach my financial goals for the year. But all this hard work was keeping me from Khaiti's farm. For years I had dreamed of becoming a full-time farmer, and I hoped eventually to grow vegetables for a living. One of my goals for the year was to find some cheap land and build a house. I had saved up a little money to do this, but now I was faced with the unexpected possibility of sharing this dream with another.

"Getting dirty" takes on a different meaning when you're farm-dating.

Khaiti's two acres were full to the brim with activity. A friend had offered me a piece of land for a reasonable price. We took a trip to scope it out and became swept up in the romantic idea of a new life working side by side, so we decided to buy the land right away. But the negotiations with my friend were vague and took long enough to make us hesitate and then withdraw from the deal. Now we started to scour the real-estate listings all over the area. Everything seemed either way too expensive or way too crappy. I wasn't even sure if we were really planning to buy land together or if we were just beginning to explore our options.

I took a day off to spend it with Khaiti, and we impulsively decided it was Pig Day, which meant we were going to go find a couple of good piglets and buy them that day, no matter what obstacles lay in our way. It was a gorgeous, early-summer day, and every leaf sparkled. The sunlight gave everything the quality of a dream. We combed through the classifieds and found a few feeder piglets for sale, but they were all too expensive.

According to one hog farmer, all the 4-H kids were buying up the piglets, so the sought-after breeds were being sold at a premium. We finally found a couple of feeders at a decent price just twenty minutes north of Osceola. We drove fast, whizzing by old box elders sprinkled in the ditches.

We found the pig farm, a miserable ramshackle place. Huge piles of junk and machinery were everywhere. The farmer told us we couldn't enter the facility because it was too dangerous for the pigs' health. A dozen large pink hogs lay in a small muddy pit north of the barn, bored and lazy-looking. He asked us what kind of pigs we wanted. Other than healthy and cute piglets, we had nary a clue as to what breeds we were in the market for, so the pig farmer suggested a couple of types and we nodded dumbly and said, "That sounds great." He ducked into his mysterious closed-off barn and emerged holding two squealing piglets by their back feet, tossed them roughly into the large dog carrier we had borrowed, and stowed the carrier in the back of the station wagon. Khaiti thrust a wad of bills at him and we jumped in the car and sped off.

There was a swine stink in the air and short squeals of piglet terror coming from the back of the car. We looked at each other with wide eyes. We were delighted, full of trepidation, and happily flustered. We would take good care of these pigs. We would give them a great life, and they in turn would provide us with a lot of enjoyment and, ultimately, food.

"I think that guy was crazy," I said, "And I guess we have pigs now."

"Yes, he was, and yes, we do! I told you it was Pig Day!" Khaiti smiled at me, and I wished fervently that every day could be Pig Day. (But then of course we'd have too many pigs.)

We brought the girls home and got them into their hog-panel paddock in the drafty hoop house. They were tense and suspicious of their new home. They also produced a prodigious amount of poop. After a few minutes, they settled themselves into some straw and stopped moving. They huffed noisily and eyed us warily. There was a chill in the air, and they had just come from a warm barn. We hoped they could adjust without too much difficulty. We relaxed a little and agreed that it felt like this little farm had just grown up. Getting a couple of pigs seemed as if we'd received a "This Is an Authentic Farm" certificate.

The summer passed quickly in a blur of excitement and growth. The pigs gave us a lot of entertainment and manure. We gradually learned the ins and outs of pig farming, and in the end they provided the most delicious pork I've ever tasted.

Throughout this whirlwind year, we made bold choices, and we didn't second-guess them. Without hesitating, we plunged forward and were rewarded with an incredible year full of hopes and dreams come true. Yes, I was worried here and there, but it seemed like the universe itself was behind the steering wheel of our crazy lives, so I let go and enjoyed the ride. With bold decisions come bold consequences, and with any luck life becomes exciting and rich.

That summer we found the farm of our dreams and bought it together. We moved there, and a few months later I realized that there was no point in putting off the inevitable. I asked Khaiti to marry me. Happily, she said yes, and we're now on an endless farm-date in west-central Wisconsin. ✍

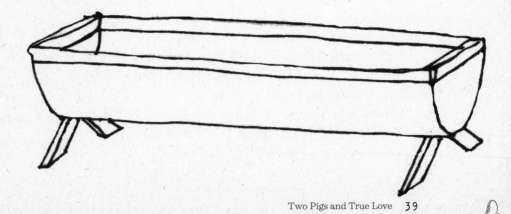

The Fruits of My Labor

BY MAUD POWELL

On their farm, Wolf Gulch, in the Applegate Valley in southwestern Oregon,
Maud Macrory Powell and her husband, Tom, grow produce and seed crops and
coordinate the Siskiyou Sustainable Cooperative CSA. Maud also works part
time as an extension agent for Oregon State University. They have two kids,
Grace (twelve), and Sam (eight).

Every year, beginning in late August, I roll up my sleeves
and give myself over to the fruits of our farm. Most of my waking
hours at home are spent in the kitchen: chopping, squeezing, slicing,
stirring, pouring, and then pulling beautiful jars from the hot-water
bath and checking to see if the lids have sealed properly. I feel trium-
phant as I carefully line up the jars on our pantry shelves. Gone are the
days of gourmet recipes for relishes and chutneys; I'm simply trying
to get as much food in sealed jars as quickly as possible. Once harvest
begins, the clock is ticking.

Every fall, I remind myself of the chocolate factory skit from "I Love
Lucy," in which Lucy and Ethel resort to desperate means to keep up with
a conveyor belt. The work is neither precious nor exalted; it simply must
be done. The heavy frosts of early November are approaching rapidly.

Meanwhile my husband, Tom, is outside transplanting fall crops,
pulling drip tape off the fields, tilling in the summer crops, and broadcasting cover-crop seed. Part of me longs to be out there with him; I'd
like to be moving my body more and feeling the elements on my skin.
But the reality is that all the work needs to be done, and no piece of it
is more valuable than another. Yet I'm still sometimes shocked at the
conventional role I've ended up in on our farm: I'm in the kitchen and
Tom is in the fields.

Tom and I first apprenticed on a farm together twelve years ago.
During our training period, he and I followed a parallel trajectory in

agriculture. We learned to drive a tractor, set up irrigation, and transplant, cultivate, and harvest alongside each other. Perhaps because I was brought up by a feminist mother who was one of a handful of women in her law school class, I believed that my career should be high-powered, professional, and linear. In the world of agriculture, that meant operating heavy equipment, making decisions about what grew in the fields, and controlling all the systems on our farm. It meant that Tom and I would split the work equally and our roles would be interchangeable.

While I was pregnant with our first child, I imagined working in the field with our baby strapped to my back, taking short breaks to breastfeed in the field. I intended to keep up with Tom and stay on equal footing in our farm operation. But toward the end of my pregnancy, I became acutely aware of the biological differences in our bodies. I had grown a human child in my uterus and would shortly be feeding her from glands in my fatty tissue. Tom had also engaged in sexual reproduction, yet his body remained fundamentally unchanged. Even so, I clung to the hope that my relationship to farming would continue to be the same as Tom's. Pregnancy, birth, and lactation were to be peripheral activities that briefly called me away from my life's purpose: farming.

Almost as soon as our daughter, Grace, was born, Tom and I fell into a traditionally gendered division of labor. I completely underestimated the amount of energy and time breastfeeding and childrearing would take. I also underestimated the love and devotion I would feel for my child, who immediately became the center of my life.

When Grace was three months old, we moved to our newly purchased farm, Wolf Gulch. Tom set to work with the determination and tirelessness of a beaver: building ponds, surveying the hilly terrain for irrigation lines and field contours, planting trees, and repairing an old barn and shed. By default, I started managing the kitchen, the baby, and the house. The trajectory of our lives suddenly split; our daily activities and tasks, which had been virtually identical, now barely overlapped. I felt torn and confused. On the one hand, I loved caring for our baby and couldn't imagine handing her over to someone else for hours a day. On the other hand, I could feel the sense of an equal partnership on the farm slipping away each day, as Tom alone learned the nuances of running our new tractor and designed complex infrastructure systems that would serve as the foundation of our farm.

For several years, I struggled to find my rightful place on the farm. I found it difficult to value my behind-the-scenes roles as CSA organizer, field helper, mother, cook, and food preserver. Four years after Grace was born, I gave birth to our son, Sam. Even though Tom and I made most decisions together, the difference in our roles rankled me. I felt a subtle yet incessant gnawing sensation: I must be the same as Tom in order to be of equal value. And yet time was always short, and we gravitated to the tasks we were best at, which meant that our gendered roles became more entrenched. It was not until we began producing vegetable seed crops eight years ago that I truly began to appreciate my role as a female on our farm.

The difference in our roles rankled me. I felt a subtle yet incessant gnawing sensation: I must be the same as Tom in order to be of equal value.

Letting a plant go to seed is actually considered to be a fairly negligent act in the world of farmers and gardeners; the very phrase "go to seed" connotes dereliction. But my favorite part of growing seeds is the opportunity to observe plants through their life cycle. I'm always awed by what happens to them when they're given free rein to manifest their biological potential. Onions and leeks send up four- to five-foot stalks with spiky balls of white or purple blossoms. A bed of parsnips transforms into a ten-foot-tall, densely matted forest of brown, feathery branches. A zucchini will grow to the size of a small ruminant; radishes become cantaloupe-size. My pregnant belly comes to mind, stretched beyond capacity and bursting with excess life. In my more philosophical moments, the produce we eat for dinner seems like lives that have been cut unnecessarily short.

Seed farming demands an intimate knowledge of a plant's mating habits. Vegetables are no longer genderless. Some plants are male or female; others have flowers that are distinctly male or female; still others have blossoms that contain both male and female plant parts. The reproductive characteristics of each species affect how much isolation is required for a plant to produce good seed. Gender is significant, and creates dynamic relationships that warrant attention and care.

One of the by-products of seed farming is usable produce that can be preserved once the seed has been extracted. Most seasons I'm left with the exhilarating challenge of putting up at least two thousand pounds of tomatoes and peppers for sauces, paste, salsa, and ketchup. I also dehydrate the flesh of dozens of melons and freeze hundreds of eggplants that have been grown for seed. By allowing the feminine gift of reproduction to flourish on our farm, I'm drawn back into my kitchen every August.

As our kids have gotten older, my life has moved into ever widening circles. I'm a social creature by nature, and have found myself in various community-organizing roles. After hosting farm interns for several years, we helped start a nonprofit devoted to improving the quality and education of farm internships. I received a grant to write a curriculum for interns, which helped launch the organization's farm education program. Rogue Farm Corps currently has fourteen participating farms and educates at least twenty-five interns a season. I also started doing work for our regional agricultural cooperative and received grants to purchase seed-cleaning equipment for the group. Then, Tom and I became the coordinators of the cooperative's multi-farm CSA program, which comprises fourteen farms and ranches. A few years after that I started working part time at Oregon State University's local extension office, teaching classes on farming.

I enjoy feeling that I have a positive impact on the larger agricultural community, but once again I feel torn. This time, I wonder if I'm spreading myself too thin and spending too much time away from the farm and the kids. How much more successful would Wolf Gulch be if I plowed all my energy into our own operation? But the truth is that

I find a deep satisfaction in working at the heart of a larger movement, on a scale greater that the perimeters of my own farm. I like to think of myself as planting seeds in the hearts and minds of my students and helping to preserve the bounty of our rural culture and economy through cooperative efforts.

Tom has remained supportive throughout. Ironically, our relationship has turned out to defy some typical gender dynamics: I'm the one who leaves home two or three days a week to work in an office while he stays on the farm. The world in which I operate is much larger than his, yet our relationship continues to draw me back into the domestic sphere, where I find a different form of contentment.

Like a faithful cat proudly bringing home dead rodents for its owner, Tom arrives at the kitchen door with bushels of tomatoes and peppers for me to process. He looks both pleased with himself and a little sheepish at the abundance of the harvest. I feign dread and even resentment because I want him to appreciate how much time and monotonous labor goes into transforming these fruits into two valuable products: seed and preserved food. Tom, who indulges me just enough, expresses his gratitude for my labor. Secretly, I'm pleased by the fecundity and fertility of our farm. I am also pleased by Tom and his efforts, and I strive to match his work ethic. In this moment, I appreciate that our distinct roles create an intimate and interdependent dynamic between us. I also feel at peace with my desire to work in both the microcosm of our farm and the larger sustainable agriculture movement.

This past August, as I checked the temperature of the canning bath and felt steam hit my face, I was struck by the completeness of my life. I'm like a plant that has been given free rein to manifest my biological potential. When the canning season wanes, on the eve of the first predicted frost, the kids and I always join Tom to gather up the winter squash and stash it in the safety of our barn. That night, after supper, we ritually open our first jar of canned peaches and light a fire in the woodstove. The peaches are always sweeter than I expect them to be. ✍

Surrender

BY COURTNEY LOWERY COWGILL

A writer, editor, and farmer based in central Montana, Courtney Lowery Cowgill is the cofounder of the online magazine New West and a columnist for the journal The Daily Yonder. She and her husband run Prairie Heritage Farm, where they raise vegetables, pastured turkeys, ancient and heritage grains, and sometimes a little ruckus.

One Friday in early September, I dragged myself out to the field to harvest for the next day's market, ready to spend hours picking, weighing, and sorting the overwhelming bounty a September harvest day brings.

Instead, when I arrived, my husband shouted from across the field four small but feared words: "We got a frost."

The forecast had called for 38°F, but in our little spot that had quickly dipped below freezing. We'd harvested the first of the tomatoes the week before and the peppers were just starting to put on fruit. It was a cool, wet summer and most of our hot- or long-season crops had been seriously hampered. We'd been waiting, hoping, for an Indian summer. Some of the winter squash — what would feed us and our customers all winter — had only recently started to flower. We still had five weeks of CSA deliveries for our customers and four weeks of farmers'-market tables to fill.

As we walked through the beds, inspecting blackened tomato leaves and wilting summer squash plants, Jacob said, "I'm trying really hard not to feel bad about this."

I felt it, too. Overwhelming defeat and, moreover, a feeling of guilt.

It's baffling to me that something beyond your control can make you feel like a failure. But welcome to the life of a farmer.

The cliché about the complaining farmer is an easy one, but it's accurate. Find a farmer who doesn't complain about the weather or wheat prices or sawfly or federal farm policies and you've found a rare species. Listen closely, though, and I think you'll hear that the complaining isn't just for its own sake. These farmers are not looking for sympathy. The griping is an acknowledgment — a plea for some sort of recognition — that all of this is out of their hands.

As a farm kid, I'm no stranger to that out-of-control feeling. I grew up on a farm that tried to play the commodity game. Wheat, barley, wheat, barley, and all of it sold for whatever the market told us to sell for. Being at the mercy of the weather and the pests and the soil was one thing. Being at the mercy of the market and the grain elevator added insult to injury.

When my husband and I started thinking about farming, we set out to do so by asserting more control. After watching my family go through what we did in the eighties and nineties, I would dive into farming only if it meant more stability on the farm (and thus in the family) than what I had known as kid.

That's how the sustainable ag/local-food movement won me. Direct, local markets meant predictable markets. No commodity markets meant we got to name the price and call the shots. Less dependence on government help meant less intrusion, direct or implied.

The first thing we did was set up a CSA for the farm. In a CSA, customers, or shareholders, as they're called, buy in up front for products, giving the farmer the capital needed to grow or raise the food and establishing a ready market. It creates a wonderful relationship between farmer and consumer. We share in the risk and we share in the bounty.

For our first CSA, we sold shares for Thanksgiving. Early in the season, shareholders paid for their heritage turkey, winter squash, potatoes, and onions — all before the turkeys were hatched and the seeds were in the ground. Our next CSA was a traditional vegetable operation in which members paid in winter for a weekly delivery of vegetables during the growing season. In our second year, we added a grain and seed CSA in which members get one big delivery in the fall of heritage and ancient grains, specialty barley, and lentils.

This is nothing like the farming I grew up doing, and that first year, with every CSA check that arrived, I relished what I thought was control. We'd figured it out, I thought.

Oh, how wrong I was.

No matter how much control our business model gives us, we are still farmers and farming is mostly an exercise in managing chaos — an attempt to control the uncontrollable. No method or scale or marketing strategy can change that.

Out on the farm, and in the bedroom, the more things I tried — lights, water, fertility charting, worrying — the less happened.

That spring, I heard that message loud and clear. In my early-planting schedule, I'd sown flats and flats of basil seed. I couldn't wait to see the little sprouts. I watered and rotated the lights and waited. As all our other seeds sprouted and moved on to the greenhouse, the little basil flats stayed brown. I turned up the heat. I watered more, I watered less. I sprayed instead of sprinkled. I replanted. And still, weeks after they were supposed to germinate, nothing.

We were trying to get pregnant at the time and fertility, or lack of it, surrounded me.

Ironically, when we first moved onto the farm, I dubbed the potting shed — where we germinate our plants — "the womb." That first spring I spent a lot of time in the womb, and every week: no germination in the greenhouse. Every month: no germination in the bedroom either. With every passing day, I felt more and more defeated.

I must be doing something wrong, I thought. I'd gotten accustomed to a modern life that taught me that if I wanted to succeed, if I wanted something, I needed to be more, say more, do more, pray more, move more. So naturally, if I wasn't getting what I wanted, there must be something more I could do to make it happen. Especially in my career in online media, I'd gotten used to making things happen instantaneously. In this modern life, I think we've all gotten used to making things happen instantaneously.

But out on the farm, and in the bedroom, the more things I tried — lights, water, fertility charting, worrying — the less happened.

Finally, frustrated, I decided to give it over — not give it up, but give it *over*. I came to terms with the fact that all of this was out of my hands. I'd done everything I could and it was time to surrender — to God, to Mother Nature, to fertility, to the basil gods, to whatever. I wasn't in control, and I had to be able to be okay with that.

Surrendering wasn't easy, but once I did it, it gave me a powerful, freeing feeling.

In the solitary days on the farm, it's easy to find metaphors for life among racing winds and sprouting plants. Some of the metaphors don't make it beyond that particular day, but this one, the one of surrender, is the one I've carried with me since.

I'm not saying I don't still complain. I do, but I realize now why I do it. And now, when I hear other farmers complain, I recognize the subtle wonder in their voices. It's less of an "oh-poor-me" tone and more of a "can-you-believe-it?" one.

The truth is, most of us know that despite our attempts at marketing, at pest or weed management, at crop rotations and fertilizer, a majority of what happens on our little patches of land is beyond our control. And that's as it should be.

After all, being out of control means we're connected to nature, to our crops, to God, to something bigger and mightier than we are. And for many of us, that's one of the main reasons we do what we do in the first place.

That spring, eventually some of the basil germinated and what did grew up tall and luscious and kept our customers and us in pesto for the winter.

About a year later, when we'd stopped "trying" for a baby, I saw two wondrous pink lines on a little stick. The next fall, we added a little farmhand.

That October, I took five-week-old Willa into the greenhouse to harvest what was left of the basil. After we had lost all of the outside basil to that September frost, aphids — fittingly — got most of what we'd planted in the greenhouse, and I thought I'd had my last taste of fresh herb. But when cooler days killed off the aphids, I was shocked to find a few resilient basil plants still putting up shoots.

I made pesto out of what was left and sometime in January, when I can't imagine something green and sprouting, I'll spread it on crackers and start teaching Willa about the power you can gain when you let go. 🌿

CHAPTER TWO
MONEY

Somewhere along the way in school
you learn about the nitrogen cycle, and that nitrogen is the big limiter
when it comes to plant growth. Later, when you become a farmer,
you realize it's true. And unless you're already a tycoon, you inevitably learn about the other limit to growth on your farm — it's called
Money. Whether you're trying to buy land, or build a barn, or afford
a tractor, or fill the tank in your delivery rig, most of the time it takes
money.

And most of the time, we farmers don't have it.

Nor do folks — be it the bank, or the government, or shoppers
at the farmers' market — always want to give it to us.

I learned that lesson the hard way during the first spring of
running my own farm. I had saved and I had penny-pinched and I
had scraped and I had scavenged in anticipation of the cash outlay
it would take to get the farm off the ground. Nevertheless, come
April I had spent everything and my first hope of income was still
two months out, growing slowly through a cold, wet spring. I was in
the midst of what they call a Cash-Flow Crisis. Sounds dire, and in
fact, it was.

I had hoped to qualify for a USDA Beginning Farmer Loan, a
federal program that until then had stirred up feelings of pride
and patriotism in me. What a great government to earmark funding specifically to support young farmers! But when I told the
loan officer that I needed the money to pay for a buried irrigation
mainline on the family land I was row-cropping, he shook his head

unapologetically. Sorry, no money for permanent improvements on leased land. No matter that it was family land that I intended to lease for a lifetime, or maybe someday own. Never mind that I was the very demographic he was purporting to serve: young, limited resource, female (a.k.a. "minority"), and just starting out. It was a bitter pill, getting a USDA slap-down in my very first season.

The banks told me they'd happily give me a loan, at 12 percent interest. And my parents didn't have any spare change to help float me until June.

In the end I resorted to using a credit card with a twelve-month 0 percent interest rate to finance my first year of farming — the scariest, most out-on-a-limb financial risk I'd ever taken. Thanks to a good enough growing season, I was able to pull myself out of credit-card debt before the 18.9 percent interest rate kicked in.

The financial stress of farming doesn't ever go away completely. We have good years and we have bad years. We have business growth spurts that require elusive capital, and we routinely weather unforeseen, expensive disasters — hurricanes, floods, droughts, barn fires, broken tie-rods, runaway horses, diesel thieves, global economic meltdowns, cucumber beetle infestations, hungry deer, and the neighbor's bloodthirsty dog. Things that can knock us flat, but hopefully don't leave us broke or broken.

These essays tell you about some of the financial woes and middle-of-the-night ulcers, as well as the perseverance and passion that won't be intimidated into surrender by loan officers, banks, parents, or bleak balance sheets. These stories admit that our farming endeavors are defined largely by money, but they also remind us that money alone cannot, will not, does not define our success.

— Zoë Bradbury

How Not to Buy a Farm

BY TERESA RETZLAFF

In 2003, Teresa Retzlaff and her partner, Packy Coleman, began farming on the north Oregon coast. Six years later they managed to purchase land near Astoria, and now live and farm on 46 North Farm in Olney, Oregon, where they're building both their soil and a very big elk fence.

"Do you really want to keep farming?"

We asked ourselves that question more often than I want to remember: each and every time we found ourselves flat on our backs, gasping for breath after hitting yet another brick wall on the path toward buying a farm.

We had reached the end of the road at the farm we'd leased for five years on Oregon's northern coast. (Lesson learned: If you're going to lease land for your farm from very nice people you know and are maybe even friends with, be sure to put everything you are agreeing to in writing. Be explicit. Then have both a lawyer and a therapist listen as everyone involved explains exactly what is being agreed to. And then still have a backup plan in case it all goes to hell.)

We were self-taught farmers, former urban folks who had followed the "just jump off the cliff and figure out how to flap hard enough to sort of fly" method of beginning a farm. With the help of family, friends, a few good books, a very effective small-business management class, and many extravagant but useful mistakes, we had built a strong local following for our herbs, flowers, and edible plant starts that we sold mostly at farmers' markets.

Still, if we wanted to really make it as a farm and reduce the amount of off-farm income required to make ends meet in the winter, we needed to expand. We had to start growing produce and fruit, keep chickens for eggs, develop some value-added products, and maybe start to offer on-farm workshops. All this would require a significant investment in infrastructure and long-term land security.

We wanted to make a serious, stable, long-term commitment to a piece of land, to plant fruit trees and blueberries and know that we would be there to harvest them when they matured. We wanted to take the time to build good, rich soil and not have to walk away from it just as it started to produce well. We were wary of leasing, knowing now what it meant to invest huge amounts of money and years of our lives into land that, in the end, you have no legal right to. Any lease can be broken. We needed to buy our own land to farm.

We found a wonderful piece of land, eighteen south-facing acres outside of Astoria, on the north Oregon coast. It was out in the country but still close enough to town to make selling practical. About half the land was well-drained, silty loam, which in that area is like striking gold. The small house was in good shape, it had a lovely old barn that desperately needed a new roof, and a couple of other outbuildings that could best be described as rustic. It was for sale by owner, and priced optimistically high to try and catch the property-market boom that was only just beginning to slow. It had been for sale well over a year.

It was spooky how much the place met our needs — it truly felt like this farm was where we were supposed to be.

Being a small farmer in the Pacific Northwest can at times lead to some insular thinking. All the positive feedback about the work we do — from customers at the farmers' markets, restaurants, local media, local food organizations, and schools — is tangible. Tell people you're a small farmer and chances are they'll beam back at you and say how great that is.

We felt good about our chosen work, and about our plans for the future.

With our parents' help, we had enough money for a decent down payment. We had a solid and realistic business plan that anticipated higher levels of off-farm income in the beginning, as we built both soil and infrastructure. Before we could negotiate for the land, however, we needed to know we could get a loan, so I started calling around to banks to talk about small-business loans. By the third or fourth bank,

I was getting used to the deafening silence that always greeted the word "farm."

"Well," a banker would say slowly, clearly aware that he was talking to a crazy person, "we don't do farm loans."

"But we're a small local business," I would counter. "We need a small-business loan. We work with growing plants, and we sell to other local businesses or directly to local customers. We need to move our small business to a new location. It's a live/work situation. Would you like to see our business plan?"

We finally found a bank that didn't flinch when I said "farm," a branch of a national bank that prided itself on its super-green, socially responsible business image. The people there were enthusiastic about local organic agriculture, loved the part of our business plan that involved making the farm available for school tours, thought our plans for wildlife habitat in the wetlands on the farm were great, and said they looked forward to working with us to bring more fresh local food into our community. The 8 to 9 percent loan they were offering made our eyes water, but we were getting desperate, so we kept talking.

It didn't help that our loan went up for review the day the stock market crashed in September 2008, although that wasn't the reason they gave for turning us down. Crazy as it seems, our business plan involved us actually living on our farm. And unless we could guarantee that 100 percent of the income we would use to pay back the loan would come solely from farming —which meant no off-farm income to cover our expenses while we built infrastructure and waited for our first harvest — what we were really asking the bank for was a home loan, which it didn't do.

I'd like to meet the farmer who can buy a piece of land, till it, prepare the soil, sow the seeds, grow the plants, harvest the crop, take the crop to market, sell it, deposit the cash in the bank, and write a check for the mortgage all in one month. In October. On the Oregon coast.

Hoping to talk to people who at least understood the physical realities of farming, we called the USDA Farm Service Agency about its small-farm loan program. The FSA agent we met with listened to our plans, then paid us the compliment of saying that it seemed like we actually had our act together. But she went on to be brutally honest about the FSA small-farm loan program and our chances of actually getting a loan to buy land for our farm. To her, a small farm was one growing four hundred acres of grass seed or running three hundred head of cattle. She told us that our proposed five or six acres of cultivated land growing mixed vegetable, fruit, and flower crops, and raising chickens with some off-farm income rounding out the economic edges fell into the USDA category of a "lifestyle farm."

"We don't normally make loans for lifestyle farms," she told us politely.

"It's a damn hard way of life, not a bloody lifestyle," I muttered, annoyed, on the subdued drive home.

The seller of the farm we loved still wanted a crazy amount of money for it, and we had no loan options that would let us even begin negotiating, so she stopped talking to us and we sadly tried to accept that we were just never going to farm that land.

We spent the next six months scrambling, trying to come up with some way to keep farming on the Oregon coast. We had market customers calling us with land-for-sale referrals and offering to sign up for a CSA program before we even had a farm to grow the food. We explored many, many ideas: buy land with a group of people and start a nonprofit education farm; temporarily lease another piece of land; find a cheaper land option; renegotiate with our current landlords; borrow money privately. Each option was explored and each gradually disintegrated as we tried to cobble together a solution to keep our dying farm alive, all painfully in the midst of our best market season ever.

A grim day in June found us sitting at the kitchen table facing the bleak reality that we were going to have to quit farming. It was a painful moment for me. At forty-three years old I had finally found work I loved, work I was actually good at and that I cared passionately about. I could grow plants, I could feed people, and I could teach them how to grow plants and feed themselves. The support from our community for the farm we wanted to have was heartening to us, but it couldn't get us the loan we needed. With deep resignation, we each made phone calls, went for interviews, and accepted "real jobs" with the understanding that we would start part time to allow us to finish out the current farmers'-market season, pay off bills, and put the farm into hibernation.

Farm or no farm, we needed to find a new place to live. While cruising around online to figure out what kind of house price we might be able afford with our new job income, we stumbled across a local real-estate company on whose home page under the heading NEW LISTINGS was that farm. The farm we loved. The farm we'd tried to buy for more than a year, the land we'd dreamed about and planned for and had finally, depressingly given up on some six months ago. Still for sale. Price reduced to something we could now maybe afford.

```
The irony of having to quit farming
so we could finally get a loan
to buy the land to move our farm to
stuck in our craw.
```

In a daze, we called a local bank and made an appointment to talk about a straight-up, super-normal home loan. We told our long story to the very nice broker, reassured him about our commitment to our regular-paycheck "real jobs," described the down-payment fund we had waiting, and explained our plan for keeping the farm going part time to help with additional income to pay the mortgage.

"No farming," he said sternly. "Quit the farming, right now. Only work the regular-paycheck real jobs. Then, maybe, we can make it work."

So that's what we did. The irony of having to quit farming so we could finally get a loan to buy the land to move our farm to stuck in our craw, and was made even harder to swallow when we had to provide written reassurance to the lenders (nervous about our worrisome "history of farming") that although we had indeed spent five years running a "hobby farm," we had seen the error of that life path, now had nice safe real jobs, and only wanted to buy eighteen acres of land zoned agriculture-forestry so we could continue to live a "rural lifestyle."

It wasn't a legally binding document, and besides, I had my fingers crossed behind my back when I signed it. I can't say I recommend lying to your bank as a road to farm ownership, but it worked, and I'm not ashamed that that's what we did. The shame I feel is for a country that makes it virtually impossible for hardworking beginning farmers — people who are willing to devote their lives to growing healthy food for their communities — to own land.

We're still working off-farm to make ends meet, slowly building our soil, rebuilding our infrastructure, putting down roots, heading back to being farmers again. Challenges are still there every day, and they always will be. Some of them seem impossible in the moment.

"Do you really want to keep farming?" we ask each other.

And the answer is always, "Hell, yes." ⋇

Worth

BY BEN JAMES

With his wife, Oona Coy, Ben James runs Town Farm in Northampton, Massachusetts.
This year he is obsessed with compost-turning pigs, greenhouses on wheels, and
doubling the value of food stamps at farmers' markets.

Last week at market a customer complained about the price
of our dill (two dollars for a not-huge bunch). He said the price was an
outrage, but he was smiling, so I was too confused to ask why he was
going ahead and buying the dill, or even how he'd arrived at his notion
of its value.

This is not an unusual occurrence; every week at market we get at
least one or two potential customers who shake their heads in dismay
at a $2.75 head of lettuce or a $4.00 pint of strawberries. Sometimes I
engage in conversation, sometimes I don't. I try not to get defensive,
and I frequently encourage a customer not to buy the product, offering suggestions of where to find cheaper food, either at the market or
elsewhere. I do my best not to reveal that the value of our produce is a
question that regularly fills me with a tremendous amount of anxiety.

What is a carrot worth? A bunch of kale? A handful of berries?
Too often, I find myself on the tractor making quick calculations in
my head. For a bed of carrots, there are the soil amendments, the
cover crop last fall, the chicken manure, the organic fertilizer, the
plowing, tilling, seeding, irrigating, thinning, weeding, harvesting,
washing, bunching, packing, and selling. Plus the cost of the tractors,
implements, and fuel. Plus the cost of childcare and preschool. Plus,
somehow, all the time spent on the computer (where does that fit in)?
And I haven't even mentioned the cost of the land (hundreds of thousands of dollars, in our case). The sheer number of labor hours and
material and property costs that went into helping this soil produce
these carrots. I ought to shellac the carrots and hang them on the wall.

For us, the value of our produce can be measured — at least imprecisely — by how hard we and our crew work to grow it.

But what if the workers were just slow weeding the carrots that day? Or what if the farmer himself is a hack? What if it takes him three seedings over that many weeks before he even manages to get a row of carrots to germinate? (I'm not naming names . . .) Should the customer be expected to pay for the incompetence of the grower?

Fortunately for me, I suppose, incompetence is much less an issue than the very nature of the project we've undertaken. We grow many different crops (forty-six and counting) on a small amount of land (eleven acres), and this — as each of our variously weedy rows can attest — is a fundamentally inefficient thing to do. Although we strategize endlessly about how to make our operation run more smoothly — setting up systems, buying new equipment, instructing and correcting the crew — it's a hopeless endeavor. Eventually, we'll need either to substantially increase the size of our farm or shift our marketing strategy to grow only a handful of the most profitable crops. Until then, we mechanize whenever and wherever we can, but even the potato harvester and the water-wheel transplanter I've got my eye on would have a hard time paying for themselves at our scale, and so we're left with that most versatile and least cost-effective of technologies: our hands.

```
The value of our produce and the
value of our labor are unsolvable
computations that I puzzle and
worry over constantly.
```

All of our hands: Oona's and mine, the hands of our four full-time employees, plus the scattered extra people who frequently fill in the week. And if there's anything to match the anxiety of assigning a value to our produce, it is, for me, the challenge of figuring out how much to pay our crew (currently at least $9 an hour).

I was raised by leftist labor organizers in Kentucky, Detroit, and Queens, and it's fair to say that the plight of the Big Boss Man was not a frequent topic of conversation around our breakfast table. I learned the importance of work and the compromised position of the worker, and I was taught to question at every level the judgment and the ethics of the person in charge. So, to that small subset of the American population that was raised in the inner-city by Marxists before going on to start small, diversified farms and employ several recent college graduates, I say, "Hey, I can relate." It's not easy to be a boss, especially when your workers are getting paid more than you are, the pigweed is as high as your navel, and the man at the farmers' market is smiling while he complains about the price of your dill.

The value of our produce and the value of our labor: These remain for me unsolvable computations that I puzzle and worry over constantly. And although some parts of Oona's and my situation are unusual, the basic equation is not: Small-scale farmers and their employees are earning nowhere near the money they should be making for the endless, all-encompassing, dangerous, exhausting work they're doing. Easy to say, but difficult to figure out what to do about it, whether you're the farmer or the customer.

A whole can of worms, these questions, and in the midst of all of it is my son, Silas. He is four this summer and recently he's made some startling revelations, namely that this is *our* farm, we *own* it, these are *our* vegetables and *our* workers, and Momma and Poppa are *in charge*. I'm pleased by his pride and sense of ownership in the farm, but I also want to laugh and say, "Yeah, we own these vegetables, but do you know what they're actually worth?"

I also cringe a bit at the entitlement that comes with the package. The few times I've seen Silas try to play boss to the crew, I've pulled him aside to say that when he learns how to do the jobs faster and better than all of the workers, he'll have earned the right to tell them what to do. (I don't say he'll also need to learn to see things from their point of view, but he will.)

Recently he and I walked across the road so he could meet the new lambs I'd put on pasture. Along the way we checked to see what crops were coming in. The first sugar snaps brought tremendous satisfaction. I'd known they were there, but for him it was like discovering a room you didn't know you had in your own house. His mouth was full of the juice and strings. We walked over a few beds and I pointed down a row.

"Go check out those," I said, and he stepped into the field.

"It's kale," he said.

"Nope."

"It's onions."

"No, look closer."

He bent down and rubbed his palms across the curly greens.

"Look in the ground," I said.

"Hey! It's carrots!"

He'd been asking about them for weeks, and now we pulled a few, cleaned them off in the wet grass, and he ate his first one of the season. It was pale orange and slender, not even as fat as a Sharpie marker. It was gone before I could blink. And I say this with all sincerity: It was worth it. ✄

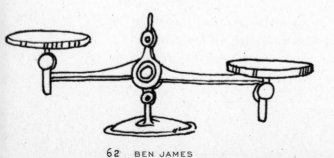

Learning to Measure Success under the Big Sky

BY ANNA JONES-CRABTREE

Along with her husband, Doug, Dr. Anna Jones-Crabtree traded twenty years of savings for 1,280 dryland acres and a diverse organic crop rotation in north-central Montana. Although honored as a Sustainability Institute Meadows Fellow and as a Presidential Sustainability Hero, Anna is most proud of her responsibilities as a land steward.

"One of the most powerful ways to influence the behavior of a system is through its purpose or goal. That's because the goal is the direction setter of the system, the definer of discrepancies that require action, the indicator of compliance, failure or success toward which the balancing feedback loops work. If the goal is defined badly, if it doesn't measure what it's supposed to measure, if it doesn't reflect the real welfare of the system, then the system can't possibly produce a desirable result. Systems, like the three wishes in the traditional fairy tale, have a terrible tendency to produce exactly and only what you ask them to produce. Be careful what you ask them to produce."

— Donella Meadows, *Thinking in Systems*

Sunday, November 8, 2009, 3 p.m. Doug and I have both crawled up into the end of "Ernie," our combine, to clean out the flax straw that seems to be packed solid in the beater grates. The quarters are tight, and Doug's a bit grumpy because we weekend farmers never get the full list checked off. We really ought to be driving home right now, since we both have big things to do at the day jobs on Monday. It's a rough time, so lately I've been working really hard at reframing our situation. At least we're out of that incessant wind for a bit and it's not freezing cold, I say. We should be celebrating.

We'll have only a few more weekends at the farm before hibernation. After all the hard work this year, all that's left is cleaning up equipment and the shop, and hauling the wheat to market. We deserve to be a bit more celebratory. But we really aren't.

Behind the scenes lurks the reality of our situation. We've both hinted at it, but haven't really found the time to dump the pieces on the table, rearrange them, hold them up to the light, and look at them holistically. We started farming at a time when prices would garner a small but positive cash flow in year two. We had a plan. We'd grow specialty crops. We'd be organically certified. We had watched the markets. We had spent years interviewing other organic farmers and attending farm tours. The five college degrees between us would really be put to use. We had savings. And as if our good credit and sheer gumption weren't enough, we'd researched and participated in all the 2008 Farm Bill beginning-farmer programs that applied to us. We often understood those requirements better than did our local USDA officials.

So often we choose measures of success based on what can be counted or quantified, and not on what our inner knowing tells us is important.

Well, guess what? The recession set in. Prices crashed. And it's rained barely three inches since June. The conservative crop yields we'd used to estimate cash flows for our business plan turned out to be bumper-crops predictions compared to what actually grew. Our equipment repair and maintenance costs were twice what we'd estimated. By all traditional measures of success, we had a terrible year. Make any money? Nope. High crop yields? Nope. Frankly, when we thought about our dream — this farm enterprise — instead of being inspired and energized, we both felt a bit worn out. And now we were hanging out in what we called the Land of the Unknown.

What really happens when you can't make any of the payments the first year? Maybe everyone else, including Farm Credit Services (the people we mistakenly thought provided credit services to farmers), were right. Dryland farm, eh? Eyebrows lifted.

But what really is success? Yes, we actually had a crop. Only seven months after signing all the loan documentation and starting virtually from scratch, we had something to harvest. In retrospect, I knew the goal of our farming adventure was never just about dollar signs. But that doesn't make it any easier to live in a society in which collectively our only goal is to make money in the short term. That's the problem we have with so many of our current systems.

And therein lies the huge challenge. Our personal farming goals — where we're striving toward diversity, resilience, ecological symbiosis, and fun, along with financial stability — don't measure up next to the larger food (that is, commodity) production system where the only goal is to make a buck.

Think about all the repercussions that stem from our hunger for the dollar. What if, as Sandra Steingraber, the acclaimed author and ecologist, has written, instead of all the economic bailouts, we actually tried to figure out what constituted an ecological bailout? I bet our five-year rotation with a minimum of ten crops (unheard of in a dryland grain system), organic production methods, and soil-building regimen might measure up better. Darn the system whose only measure is successful economics.

So we'll have a chat with our beginning-farmer loan officer. We'll grin and bear the conversation about how we didn't measure up based on traditional gauges of financial success. Maybe we'll get a loan deferment. Maybe we'll ask for more support from our already giving-tree family. They understand that the numbers aren't always what are important.

So often we choose measures of success based on what can be counted or quantified, and not on what our inner knowing tells us is important. I know deep down that we'll be able to make it work one more year. We both have off-farm salaries and health insurance. Most people would be satisfied with that life alone.

We signed up as beginning farmers right before both of us turned forty. We didn't want to wait this long to start. We tried multiple beginning-farmer/retiring-farmer match programs. The agreements were insular at best: "Work forever without management input and maybe we'll sell some of the farm business to you."

Instead, we crafted a plan to start on our own. We turned our retirement — twenty years of savings and good credit — into more than a million dollars of debt. We bought twelve hundred and eighty acres of wide-open land under the big sky of the Northern Plains. (Never mind that the closest land that could produce a cash flow was 480 miles round-trip from the "paying" jobs.) We procured from all ends of the earth a fleet of aged iron to which we've assigned names and personalities.

All this, to get to today, pulling flax straw one gloved handful at a time, out of Ernie. I say to Doug: "Maybe next year we need to find a way to collect this stuff. It'd be great cob-building material for the farm cabin." He laughs and says my brain is hilarious — it's always thinking.

I smile and say: "This is only season number one." And then I think about all we invested to get to today — our time, our financial well-being, basically our whole selves — for the opportunity to grow healthy food and model some new ideas about true working-lands stewardship.

In return we got to spend a lot of time outside, watching the raptors hunt mice after our tillage. We both lost weight. We found a great vet after the two young terriers each took an end of a porcupine in the middle of barley harvest. We saw full fields of peas and flax in bloom. We absorbed stories from our eighty-some-year-old neighbor about his father planting flax. We learned you must chain down the disk drill when using the transport. We took one step on a long journey of building, not depleting, our soil. We discovered the collective support of lots of others in the world who share a vision of agriculture as being part of and not separate from a fully functioning ecological system. And we learned that we have a lot more to learn. ✑

In Praise of Off-Farm Employment

BY CASEY O'LEARY

The owner of Earthly Delights Farm, a human- and bicycle-powered urban farm in Boise, Casey O'Leary also runs Earthly Delights Sustainable Landscaping, which focuses on native, drought-tolerant, and edible landscapes. She is writing an erotic novel set on a farm in Oregon. Yee-haw!

I didn't grow up farming. I found it through a long snake of a city road that began when I was a latchkey kid learning life lessons from Zach Morris and Jessi Spano on *Saved by the Bell* and brawling with my brother over the last bowl of Fruity Pebbles. That road wound me through suicidal depression, dropping out of college, drug-laced hippie festivals, Green Party activism, radical feminism, and finally, thank God, to Wendell Berry. At the height of my man-hater phase, I met Marty, who, despite his unfortunate genitalia, abhorred the establishment as much as I did, and together we birthed our farm as a big, giant F-You to The Man. And so we embarked on, in Marty's words, "the greatest, grandest adventure of our lives."

Throughout the last six seasons, my life has grown richer, my spirit more rooted, and my mind more creatively engaged than I ever thought possible. I've had the pleasure of meeting many other people who, by equally circuitous roads, found themselves drawn to the demanding and rewarding profession of sustainable small-acreage farming and ranching. I have a friend who left a high-paying fashion design job in L.A. to farm, another friend whose youthful train-hopping days are now memorialized in tattoos down her arms as she tends her diversified cut-flower farm. Marty's path took him through a Mormon upbringing, a stint in the Navy, a job teaching English in Korea, and a short real-estate career in Phoenix.

Although two such volatile souls as he and I are could make a go of farming together for just two seasons, we've both continued to run successful farms separately ever since. My new farming partner, Lori, left her job as a second-grade teacher (her second career) to farm.

Regardless of the individual paths we've all taken to get where we are, the thing we have in common is that we run our farms while simultaneously holding down other jobs to make ends meet.

If you had asked me several years ago, I would have been ashamed that I had another job besides my farm. I would have told you I wasn't a "real" farmer, cringing at the words hobby farmer and, worse, gentleman farmer. I marveled at articles I read in magazines about young farmers making a good living, supporting families even, from their small organic farms. What was I doing wrong? A survey of even our cherished mentor farmers and ranchers in and around Boise, Idaho, confirmed that I was not alone. Not a single one was making his or her sole income off the farm. All had other jobs or partners who earned off-farm income. Additional insight came when I began to notice themes in the locations of these miracle farms I was reading about: East Coast, southern Oregon, northern California — areas either heavily populated with millions of potential customers nearby or in the most progressive pockets in the nation. Places that leftie Boiseans jokingly refer to as Mecca.

The very real truth began to sink in: Our city, like countless other communities across America, demanded that we small farmers play by a different set of rules. After all, we were already proudly acknowledging that every plot of land is unique, each has its own challenges and abundance. Why should our farm models be any different?

The most obvious challenge I identified in our city, which may parallel your city, is that there's not enough demand for local food to support even the tiny number of people who'd like to make a living growing it. Boise can feel like a sprawling, box store–lined pit-stop along a good-ol'-boy freeway of right-to-work Wal-Mart worshippers. Or maybe, in kinder words, we're slow to catch on. Buying local food isn't hip here yet. It isn't even on most folks' radar, so we small farmers compete for this tiny, shiny, enlightened segment of our population — people who are largely ostracized by the rest of the city — simultaneously trying to educate the public in the hopes of creating a bigger market. And we do this while farming and holding down other part-time or full-time jobs to keep our farms alive.

More subtly, land in the city is disproportionately priced. You can't pay a rent or make a mortgage payment on a piece of land with the money you can make selling the food you grew on it, which means that any land used to grow food will require additional income to pay for it. So — and this is a huge so — whether or not a person is farming on land she already owns is a big factor in how her farm will look. Young farmers with secure land and capital will choose farm models different from those without. In a culture in which honesty about resources is in short supply, this honest difference is something rarely discussed in glossy-magazine articles about sexy young farmers, and can easily send a land-insecure farmer into comparative self-doubt.

In the interest of full disclosure, I took a long, hard look at the model I had built for myself that would enable me to farm while affording to live in my hometown, near my family. I farm small urban plots near my one-third-acre homestead, part time, transporting myself and my produce by bicycle. Lori and I train six apprentices a year on how to run a part-time urban CSA farm. We offer on-farm workshops and tours, speak to public schools and community groups about food and small-scale sustainable agriculture. We visit other, more progressive places and attempt to bring the wisdom of their models back to our slow-to-catch-up corner of the earth.

The rest of the time, I run a sustainable landscaping business where I have the opportunity to educate people about organic gardening, permaculture techniques, and responsible land use, right in their own backyards. The more I examined it, the more grateful for my other part-time job I became.

In two days a week of landscaping, I earn enough income to fill in the gaps in my farm's economic self-sufficiency. Over the years, these two seemingly separate businesses have melded more and more to create a unified whole that is not unlike the many disparate but interconnected elements on my farm: vegetables, fruits, herbs, chickens, bees, flowers, seeds. This partnership has proved useful for several other reasons as well.

Crossover clientele. My landscaping business reaches potential CSA subscribers and the CSA often sparks people's interest in better utilizing their own yards. The fact that I do each of these things gives me credibility in both — that is, landscaping clients seek me out because I'm a farmer and vice versa. Marty achieves this through part-time work at restaurants and doing flower delivery, opening him up to additional markets for his produce and flowers, in addition to doing some residential home gardening. My friend who left the fashion industry to farm does it by making farm bags from reclaimed fabrics and selling them to farms and health food stores whose managers love knowing that they're supporting a farmer while getting a professional-quality product their customers love. We all use our unique skills on and off our farms to weave our livelihoods.

Stability through diversification. Greater diversity in income streams provides greater resilience during tough times.

More opportunities for education. Although our farm trains just six apprentices a year, I'm reaching dozens more through my landscaping business. And more than anything, it's education we need in order to create more demand for local food in our city.

Long-term sustainability. I live an abundant but thrifty life, as do all small farmers. With the substantial additional income from my off-farm employment, I'm able to save money for accidents, retirement, and possibly a piece of farmland of my own.

I used to believe I wasn't a "real" farmer because I didn't survive on my farm income alone. But looking at it now, beginning my seventh season, I realize that the fact that my farm has been able to make it through the past six while so many others have sprouted and died around me — precisely because I haven't relied on it for my sole income — makes me feel proud. I didn't quit farming because I couldn't make a full-time living at it. Instead, I embraced the part-time nature of it, then watched my farm income go from around four dollars an hour to about nine dollars an hour by actually walking away from the farm at the end of the day, instead of pouring every ounce of my heart, soul, and energy into it. My relationships with my lover, friends, and family have improved because of my ability to keep the farm in a part-time box. I work an average of twenty-five hours a week on my farm — much more in spring, much less in winter — which is a new guilt-ridden challenge to admit. Here goes: "Yes, I don't work at it every waking hour of every day and I'm still a 'real' farmer."

My relationships with my lover, friends, and family have improved because of my ability to keep my farm in a part-time box.

I wiggled around to make room for everything I love, including people and activities away from my farm, and my business has become more interconnected because of it. My farm has touched the lives of hundreds of people in the community over the years, many of whom grew up just like me and who have found local food through outreach that I or my fellow farmers have done. Now that I've assumed the role of "successful farmer" in my community, I make a point of sharing this personal brainstorm and having this conversation with anyone I know who's trying to make it here in Idaho.

Josh Volk, consultant to small farmers, contributing writer to *Growing for Market*, and owner of Slow Hand Farm, inspired me last year when I met him at a Farmer-to-Farmer conference. He actually bragged about being able to run his farm on two days a week, touting the viability of the model for a wide variety of people. I agree with him that this part-farmer, part-X model is a crucial one for the future of sustainable agriculture, especially in urban areas. It can provide the wholesome satisfaction and greater reverence and responsibility that come from an agrarian lifestyle, still leaving room for the economic realities individuals may face. The last thing we in the cult of organic agriculture want to do is to discourage more people from becoming farmers, from learning the intricacies and crucial skills of this vital profession, simply because their desire or situation doesn't allow them to make a full-time living from their farms.

So what's a "real" farmer? I offer this definition: someone who's scrappy, who can piece together a satisfying livelihood while lovingly tending a plot of land; someone who advocates for a sustainable, food-based economy and who actively works to create such a thing within the greater context of her own unique circumstances and place on earth; someone who can wiggle. Someone who *loves* to wiggle. ✍

Fear of Debt: Should I Finance My Dream?

BY COURTNEY LOWERY COWGILL

A writer, editor, and farmer based in central Montana, Courtney Lowery Cowgill is the cofounder of the online magazine *New West* and a columnist for the journal The Daily Yonder. She and her husband run Prairie Heritage Farm, where they raise vegetables, pastured turkeys, ancient and heritage grains, and sometimes a little ruckus.

As a farmer, most of my days are filled with the tangible: the green shoots of seedlings, the softness of a tomato ready to be harvested, the cackle of a flock of turkeys roosting where they shouldn't.

But farming is as much about the intangible — things like cash flow and balance sheets. It's about your dreams, and when to take on debt to finance them. So, when my husband and I were first planning our farm, we were sure to make the cash flow just as much a priority as the crop rotation.

We were watching the local-food movement sweep across the country, and we were inspired. We thought such an idea could work at home, too. So, two years ago, we packed up our life in Missoula, I quit my job as the editor of the online magazine *New West,* and we moved back home to rural Montana to farm.

We had no money and no assets to leverage. What we did have was a whole lot of people pulling for us. We asked a handful of friends and family for a loan. We said we would pay them back in three years with 3 percent interest. It was more than they could earn in, say, a CD, and we would be paying less for the loan than we would if we'd gone to a bank. Plus, we would have a group of investors to advise us, support us, and cheer us on.

We started with a community-supported agriculture model, which means we have sixty people "subscribe" to our harvest: They pay us in

I grew up on a wheat-and-barley
farm that trudged through the farm
crisis of the 1980s. For a majority
of my childhood, the kitchen table
was buried in bills.

the spring for a weekly share of the bounty. That works well, because it gives us start-up capital each spring for seed, poults, irrigation equipment, and the like.

For two years this system has worked well. Investors are happy. CSA shareholders are happy. We've been doing just dandy as community-supported farmers.

Now, though, we'd like to expand our grain production and move toward living on the land we're farming. (Commuting to a farm just doesn't work.) But we need equipment. And we will have to buy a piece of land instead of leasing.

It all makes financial sense. The business plan works. But it works only if we go to the bank.

That just terrifies me.

I grew up on a wheat-and-barley farm that trudged through the farm crisis of the 1980s. For a majority of my childhood, the kitchen table was buried in bills. My brother and I knew to get as far away from the house as possible if my parents were in the middle of "doing books." The pressure of debt — of equipment loans, of operating loans, of land payments — was palpable in my family and in every other family like mine spread across the American prairie. But to make it in the commodity-driven agricultural economy, you had to get bigger. And to get bigger, you borrowed.

We watched farms around us — mostly the small ones, but some big ones, too — crumble under that pressure.

The 1980s farm crisis was a good example of how dangerous too much, or the wrong kind of, debt can be. But it's also true that if it weren't for the banks that line the main streets of farm communities across the country, there wouldn't be many farmers on the land at all. In a perfect world, we wouldn't need an institutional loan to get our farm off the ground. But I've watched too many entrepreneurs try to eke out a living and fail, partly because they were undercapitalized.

So, although I'm terrified of taking on debt, I know the other route can be just as risky. We recently sat down with our neighborhood agricultural-loan officer. He's just about the nicest guy in town, but isn't afraid to tell us how bad the situation could be if things don't go as planned.

As we walked through those worst-case scenarios, I realized that it's not the debt that's scaring me; it's what the debt means.

It's been easy for me during the last two years to think of our farm as an experiment, something we could walk away from. We're young, educated, experienced people. Surely we could find something else if this failed.

In fact, in convincing me to farm years ago, my husband used just that argument to reassure me that we would not end up where my parents were when my childhood farm faltered.

I've held an escape route in the forefront of my mind, though. That's what gave me permission to follow this folly and enabled me to take all the small steps we've taken thus far. It has also, however, kept me from taking big steps, and that's no way to start a business — or a life, for that matter.

It's true that each little decision — each seed put in the ground and each new customer signed up — marked another small commitment. And two years later, those have added up to one big commitment. But if all we ever take are those small risks, we're never going to get where we truly want to go.

That's why, when we were in the banker's office this month, I had to remind myself that whether or not we take on this debt, we're putting everything on the line — our hearts, our souls, our energy, our time, our family, our livelihood. These loans would just be a conduit for the big leap.

And maybe it's time we leapt. ✍

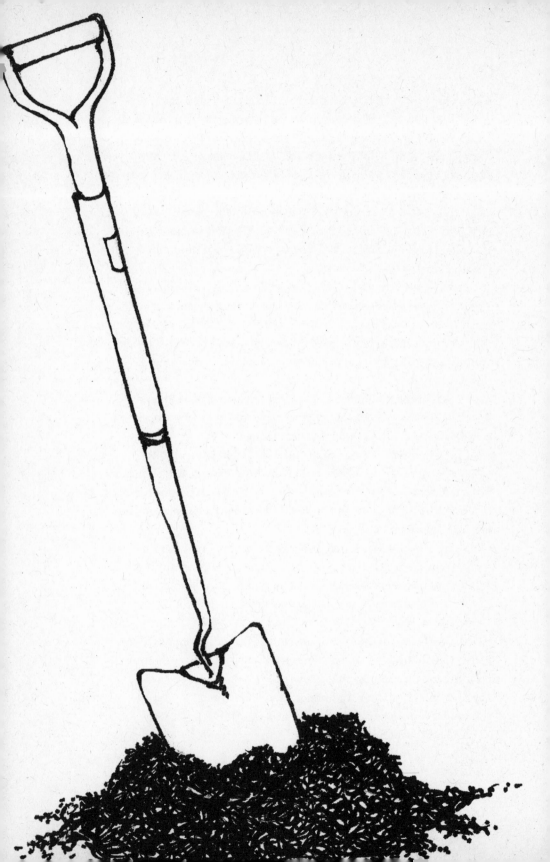

CHAPTER THREE

LAND

Beg, borrow, steal.

Behind every farm there's a story of how the farmers got their
land, and it's probably a story worth retelling. For those of us who
don't have land, figuring out how we'll manage to get it becomes
quite an existential topic. Land access, of course, is a prerequisite to
the practice of farming. And the practice of farming has tremendous
appeal, even just the act of practicing farming on someone else's
farm, as an apprentice or farm manager. Short-term access to nature,
to interaction with plants and animals over the course of one season,
only whets the appetite for longer-term access.

Stewardship satisfies a humane impulse, and the stewardship
motive swells with the prospect of watching a tree take root, bud out,
bear fruit, cast shade. Spending time in the shade of trees planted
by our farm mentors in their youth, in a well-worn, well-loved landscape shaped by the lives lived there — shaped by decades of attention
and careful thinking — we come to appreciate what a tremendous
power we have to affect the space we inhabit, to plant the rose arbor
under which our children will walk.

That opportunity — to build soil, to build infrastructure, to produce good healthy food enjoyed and valued by neighbors, to be
sovereign designer of barns and pastures — what a life! Sovereignty.
It's a concept made famous by the Landless Peasant Movement in
Brazil but also exemplified in smallholder farm culture in the United
Kingdom. I was given a little dish from Wales with this verse painted
carefully on it:

> *Let the mighty and great loll in splendor and state,*
> *I envy them not, I declare it;*
> *I eat my own ham, my chicken and lamb;*

I shear my own fleece, and I wear it;
I've lands and I've bowers, I've fields and I've flowers
The Lark is my daily alarmer
So ye jolly boys now, who delight in the plough
Long life, and good health to the farmer.

But land is expensive and farming isn't terribly lucrative, so there's some simple math to overcome. If you aren't born onto a farm, or married into one, finding land to farm can occupy most of your twenties and thirties. Some people spend their working lives in another career and then retire into farming. Usually, there's a combination of farm loans, savings, off-farm jobs, mortgages, investors, family support, partnership, and hustle. Usually, it is attended by lots of paperwork and committee meetings, endless schmoozing, and essays for the land-trust newsletter. It always requires patience.

In buying that land, we're in competition with second-home buyers and established commodity growers with a strong price-point to show the bank. Chances are the barns will need major repair. If it's a bankrupted dairy farm we're trying to buy, chances are that deferred maintenance is only the beginning of our troubles — there may also be water-quality standards, dry rot, barrels of used motor oil. (Thankfully, these days most of the junk has value as scrap metal.) Drama drama, hassle hassle: and yet we yearn for it.

Logistics aside, it's important that I set down on paper an observation born out of many hundreds of conversations with young farmers. No one will tell you this in a business-planning class, but it turns out that a surprising majority of farmers who have "landed" got that way through pure magic: "pure magic" meaning luck, serendipity, an angel investor, happenstance, or other special circumstances. Karma cannot be overemphasized. If you're doing good work, getting strong, being brave, and taking risks, a kind of magical mojo takes hold. Sometimes you get burned, but there is, in the experience of many young farmers (myself included), a particular kind of magic bubbling up at critical times that makes possible things that seemed profoundly otherwise. This oft-quoted line is credited to Goethe: "Whatever you can do, or dream you can do, begin it. Boldness has genius, power, and magic in it." Give it a try.

— Severine von Tscharner Fleming

Landing Permanency/ A Permanent Landing

BY JACOB COWGILL

A native of rural Montana, Jacob Cowgill and his wife farm fifteen leased acres near Conrad, which has brought them home and closer to their goal of owning a diversified family farm. They raise heritage-breed turkeys and vegetables, as well as grains and seeds, such as ancient wheat and milk thistle.

Early last season I stood in the field where we were about to start building a high tunnel and I wondered how we were going to get water to the structure. We knew it would require buried irrigation lines below the frost line — six feet deep, in our part of the world. Standing on leased ground, I realized we were about to spend money on infrastructure we ultimately wouldn't own.

A few days later, I watched as a backhoe clawed a hundred-foot-long by six-foot-deep trench into the ground. Irrigation pipe was laid, a couple of hydrants hooked up, and the long scar covered up within an hour. After a few days, an invoice for twelve hundred dollars arrived in the mail and all I could do was swallow hard, cut the check, and hope that it was worth it.

Just over two years ago, my wife and I were given the opportunity to start our own farm, close to where we both were born and raised in north-central Montana, on the short-grass prairie where the plains collide with the eastern Rocky Mountain Front. At the time, that's all it was — an opportunity, and one not to be taken lightly or questioned too much. Though we had to live in town, a few miles from the farm, we were thankful for the chance. Over time, we faced many of the same obstacles as other beginning farmers with limited resources, including the ability of impermanence to make every decision an agonizing one.

I used to think that finding land to farm wasn't the obstacle people made it out to be. There's land everywhere, and there are owners willing to let you farm it. And I still believe that's true, especially if you're willing to squat for the rest of your farming career. I also used to think that owning land wasn't the be-all and end-all, and to a certain extent I still believe that, too. But now I also believe that there's true value in ownership: a plot of your own, a piece of land that you can nurture, build health and wealth on, and steward for your short life, perhaps passing it on to your children or another young, land-based entrepreneur. Finding land to land on is the true hurdle.

Because of our impermanence, our desire to live where we farmed, and our intent to develop the proper infrastructure, we pushed for a plan to purchase a portion of the land we were farming. We pushed hard enough that we drew out from the farmer what he and his wife were willing to do and what they weren't willing to do. Also in pushing, we dug up the dirty details of buying land and doing so creatively.

We realized, for better or for worse, that long-term leases or creative (and complicated) ownership arrangements may be the only way for some farmers to commit to one small part of this earth. Subsequently, we came tantalizingly close to taking on debt for bare ground with no infrastructure and a fledgling business to defend. In our rush to own, without yet fully understanding the business of farming and our own unique enterprises, we lost track of just what we were trying to achieve.

After getting that we were trying to fit a round peg into a square hole, my perspective returned to that pre-farm time when the possibilities seemed endless and my farm vision was clear in my head. Now, that perspective has become colored with a feeling of deep disappointment that all the heart and soul we expended on this small piece of land may just blow away in the strong west wind if and when we move on. I worry that I may find it difficult to distinguish the short-term needs of the farm without knowing just what our long-term trajectory is. If we do end up leaving this place, I fear it will be a lost opportunity, perhaps the last opportunity.

Maybe the fact that this is our first shot has clouded our long-term vision. Instead, we should focus on the farm business, on making our enterprises as strong as possible, and not fret about buying a farm or even living on the farm. The farming should drive the farm; the land shouldn't. It's time to put our heads down, work our tails off, and hope that that is what will ultimately determine our success or failure. We need to farm for the long term, despite the short-term impermanence.

Of course, what I feel tomorrow may be 180 degrees from what I feel today. Regardless, seeds have to be sown in the greenhouse and in the cold spring ground. The organic-certification deadline isn't waiting around, and in a month, 150 day-old heritage-breed turkeys aren't going to care about my internal struggles. Ultimately, we live in (and must deal with) only the present. We can merely offer our hopes and dreams to the future. ⌀

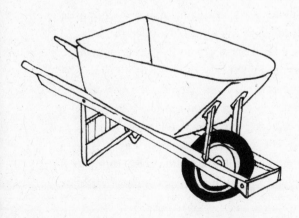

I Figured We'd Buy a Small Piece of Land

BY LUKE DEIKIS

With Cara Fraver, Luke Deikis runs Quincy Farm in Easton, New York. After eleven long months, they finally closed on their farm in spring 2011. They sell high-quality, organically grown veggies at three great markets, and have plans to expand.

Initially, I figured we'd buy a small piece of land in the conventional way — real-estate agents, mortgage, all that. I'd continue my freelance work for the film industry; my partner, Cara, would waitress; and we'd teach ourselves to farm. We'd go to a couple of farmers' markets, establish a local CSA, eventually haul produce to the city. Why not? We're both hard workers and we had a garden so lush that our friends were calling our Quincy Street apartment "Quincy Farm."

I spent all summer looking for land in an ever-increasing radius. We wanted fifteen tillable acres with a house for under three hundred thousand dollars. We were committed to finding a place big enough to grow into rather than out of. Looking back, I can't even remember how fifteen acres became the magic number. The three hundred thousand was based on some online mortgage calculator and an optimistic view of our future off-farm earnings. We blindly assumed the farm would eventually be able to take over the expenses. We started a "business plan" that today makes me a bit queasy.

I combed the Internet obsessively, cross-referencing real-estate listings to the National Resource Conservation Service (NRCS) soil maps. I'd ride out on my motorcycle to stalk possible properties, peering through the windows of houses or letting myself in through unlocked doors, poking around barns and sampling fields, trying to envision each place as Our Farm. Thankfully, some combination of the inflated real-estate prices and our small amount of good judgment kept us from pulling the trigger on any of those properties. It would have been like teaching yourself how to swim by jumping into a whirl-pool with an anvil in your arms.

The new year found us still in the city, just beginning to grasp how difficult this search would be. We struggled to expand our business plan from a bad joke to something we could actually rely on, but everything we learned shed light on a hundred things we hadn't real-ized we didn't know. Rather abruptly, we changed tactics and decided to apprentice. This buying of property was obviously going to take awhile, so why not learn how to farm in the meantime? We looked into where we'd learn the most and somehow talked our way into an apprenticeship at a great farm. As a bonus, the people there were flex-ible enough to let me go back into the city for a day of freelancing here and there. In March, we boxed up all of our stuff, said good-bye to our friends and beautiful garden, and drove away from any chance of ever living in Brooklyn again.

Apprenticing for good farmers who are committed to passing on their knowledge was probably the best choice we've made in this whole process. As we learned more and more about farming, we continued to search for land. We also gained a clearer grasp of what exactly we were looking for, and of the options out there for finding it. For brevity, I'll let three years of our life blur into one sentence: Year one faded to year two faded to year three as we left the first farm, spent a year working for another grower, then returned to the place where we'd started. The whole time we were looking, looking, looking.

We'd long since given up the concept of traditional land purchase — that quaint idea of seeing a listing, an agent gives you a tour, you make an offer, you do some negotiating, and then leverage your entire life into a mortgage. First of all, when you're looking at farmland in the Hudson Valley, you're competing with Joe Finance from the city, not other farmers. Joe Finance wants a weekend home to keep his pony, and he has big savings, big equity, and a large, steady, predictable income to qualify for that inflated mortgage. No bank in the world is going to lend that much to us — even with our excellent credit and sizable down payment. Second, even if some bank were that bold, we'd never be able to pay off the debt of a six- or even seven-figure mortgage.

We gave up the American Dream of owning land, instead separating it into two complementary goals: affordable long-term tenure for the farm and equity for ourselves. We'd schooled ourselves on alternative land tenure and could hold our own in conversations full of jargon and acronyms that would make a real-estate lawyer woozy. Our goal became forty good, tillable acres with room for infrastructure, plus a good irrigation source. Everything else was negotiable — we'd live in the greenhouse and burn sheep dung for fuel if it came down to it.

We wiggled our fingers into every nook and cranny of the alternative land-access scene: registered with Farm Link in multiple states, posted ads online with the Northeast Organic Farming Association (NOFA) and the Pennsylvania Association for Sustainable Agriculture (PASA), contacted every land-preservation organization we could track down, forced ourselves into sit-down meetings with the ones too polite to turn us away. We crafted typewritten letters and mailed them, put out the word to everyone we knew, from bartenders to CSA members. We put up a web page selling ourselves and our search, and baited it with words to catch some needle-in-a-haystack Googling landowner with fifty prime acres and a commitment to ground leases. We even perused Craigslist regularly. The game was long-term leases, and we were in it to win.

Meeting landowners and seeing their land is like being set up on a blind date with someone's son . . . someone's unemployed, somewhat homely son. These are smart, successful people taking time out of their lives and showing immense generosity by offering something of what they, blinded by parental love, think of as having incredible value — their land, sometimes hundreds of acres of it — sometimes for an extremely long term, and often for next to nothing. After a short first date we sigh and say, "I'm sorry, but I just don't think things are going to work out between us. It's not you, it's me. Really."

The first time we went to see land through a linking program, Cara and I spent several hours sitting in a beautifully remodeled living room talking about our farming experience, what kind of farm we wanted, what relationship we wanted with a landowner, our business plan, the weather, and so on. Then we went on a walk of the property, which proved to be mostly heavily wooded and very hilly. Where it wasn't wooded, it was wet. It might have made a nice homestead, but it was a far cry from a vegetable farm. We learned a key lesson: Walk the land before having the conversation.

We followed that rule for the second meeting, with a nice couple who had a beautiful, sprawling property that was just begging for an animal operation . . . but not a vegetable farm. Alas. At least we knew where we stood as we sat down for that conversation.

We quickly learned to get an address and a good description of the lay of the property before scheduling a meeting. After consulting online soil maps and satellite imagery, you can do a drive-by and, if warranted, walk the place yourself. Then, you meet.

`Meeting landowners and seeing their`
`land is like being set up on a`
`blind date with someone's unemployed,`
`somewhat homely son.`

We were amazed by how reluctant people are, though, to share this information. Many won't give you a solid answer about which way the property runs, so even if you know the address, you can't tell if it's that beautiful bottomland south of the road or that nasty wooded ridge that runs north. They wanted to preserve some level of privacy until they met us, but it sure wasted a lot of time. There were even a few people over the years who refused to give us an address until the night before our meeting, at which time they gave us turn-by-turn driving directions. Two of three properties could have been dismissed out of hand by doing research online, if we'd only known what parcel we were talking about.

The most difficult land-search experiences, though, were the ones that were almost right. Cara and I have by now spent an incredible amount of time whittling down our goal, cutting off the fat and sinew until it's a lean little sprite of a dream. We know exactly what we want. Despite that, there were times that tested us: We met a woman with six acres of good soil, a heated (if aging) greenhouse, a small tractor, a big barn, miscellaneous implements, and a grandfathered-in retail space in prime lower Hudson Valley territory. We knew we'd outgrow it in two years, but we could rock that place from day one if we said yes. Another property had a cute little house; huge, well-kept, concrete-floored equipment barns; a mechanic's garage with a pit (!); but only a few lonely acres of good ground. A tiny trickle of a creek as the only irrigation source sealed the fate on that one, but Cara and I were tempted.

Maybe worst of all was a call we got through the amorphous social network of connections we'd been cultivating. It sounded like a bust, at first — another person who wouldn't give us a description of the land or even an address — but it was so close to home that we went to meet him anyway. The landowner turned out to be a smart, outgoing film professional with a hundred-plus-acre farm. He was turning an old Dutch barn into the nicest house I've ever been in and was looking for an ambitious couple to manage the land. The property had miles of new electric fence and another attractive house that would be available to whoever moved in. But it was all rolling pasture, with only an acre or two of good veggie ground. We really considered changing direction entirely for that one, just buying some lambs and going for it, or doing greenhouses of high-end micro greens. It could have been a good life. But we held firm.

As year three sank into the prolonged ache that is midsummer on a vegetable farm, I began to revisit my old friend the Multiple Listing Service. Since we'd left the city, the economy had collapsed, taking real-estate values with it, and the boundary marking Too Far was gradually creeping farther out. One day in late June, we did a drive-by of three properties from the MLS. Two of the properties were discards and the third was the classic ambiguity: an address where one side of the street was beautiful prime bottomland and the other was mostly hilly and wooded. When we arrived, it was clear that the house and barns sat on the "wrong" side. The "right" side was so good, though, that I went ahead and e-mailed the real-estate agent representing the property. She e-mailed back several pdfs of the property lines. With delight, we realized these forty-eight acres held the house, barn, two sheds, forty tillable acres, and fifteen hundred feet of river frontage. The family selling the property had owned it since 1774, which could make us the first new family to own and farm the property since before the Revolutionary War.

As I write this, it's been six months since Cara and I first visited that farm. We've choreographed a delicate dance among ourselves, two nonprofits, and the family selling the property. One organization bought the property from the sellers, a second is buying the development rights from the first, and we'll end up owning the farm with an agriculture-friendly restriction against future development.

Working with nonprofits has made it a lengthy process, but it's going well, and with some luck we ought to close in time to move in and be planting this spring. If I could have thought of a single other occupation I'd find this challenging and satisfying, I might have walked away from farming long ago. Instead, with secure, long-term access to quality farmland, we're one step closer to achieving our dream. It's been a long time coming, but make no mistake: Quincy Farm is winning. ✑

Time on the Farm

BY BEN JAMES

With his wife, Oona Coy, Ben James runs Town Farm in Northampton, Massachusetts. This year he is obsessed with compost-turning pigs, greenhouses on wheels, and doubling the value of food stamps at farmers' markets.

I didn't notice the marks on the John Deere until I'd had the tractor for maybe a month. A couple of spots of brown, corroded metal etched into the green enamel on the top surface of the right fender. No big deal — a decades-old tractor should have all sorts of dents and dings if it's been used for anything worthwhile — but the placement of these marks was interesting. Again and again, the times I noticed the marks was when I turned around to see the row behind me and placed my hand exactly upon them, the base of my palm on the larger spot, the tips of my fingers on the smaller. I can't remember the moment now, but at some point while driving the length of one or another three-hundred-foot row I finally got it: The marks were made by the hand of the previous owner. Every time he'd turned around to check his depth or adjust his steering or see the work he'd accomplished, he'd placed his palm on this same section of fender — an unconscious action that he must have repeated several hundred thousand times — and gradually his sweat had eaten through the paint and begun to corrode the metal.

Repetition is what we do here at the farm — animal chores morning and evening, picking the squash every day and a half, the beans and tomatoes every three days, the market and CSA pickups each week, spraying the foliar fertilizer every two weeks, the big cleanout of the goat barn and the garlic harvest once a year, the three- and four-year rotations of the various families of crops — all these different time signatures in sync and then crashing against one another. There's no way to do all that's scheduled in a single day.

The tasks are completed in order, one after the other as time allows, but meanwhile there are all the ways that time on the farm overlaps and twines around itself.

We keep a log of all the things we intend to do differently next year (plant the first sweet corn a week earlier, don't follow winter squash with carrots because of the weeds, don't use biodegradable plastic under cantaloupe because the melons will rot). At the same time, wherever we are on the farm, we're reminded of what was in the ground last year (volunteer cherry tomatoes, leftover potatoes, a weedy patch where the lamb's quarters got out of control).

There's a very real sense, then, that we're farming three years at once, tracking where we've been and calculating where we're headed, even as we try to figure out — at this exact moment — what in the world to do with the eight hundred pounds of eggplant ripening in the field. (And this doesn't take into account the perennials — the straw-berries and asparagus and fruit trees — whose time signatures add an even greater level of complexity to the score).

We push and pull at time. Time pushes and pulls at us. We encourage the arugula to mature more quickly by laying down row cover, and then we load the freshly harvested broccoli into the cooler to make it last as long as it can. The bean pick we thought would take an hour ends up taking three but yields half the crop it did a few days before. The turkey that fit in the palm of my hand six weeks ago now can hardly be held in my arms. I try to squeeze a nickel out of a minute with each pint of cherry tomatoes I sell, but here's what will ultimately last: the flavor of those tomatoes in my sons' memories, so that even as grown men no other food will ever taste as good.

Time on the farm is not static, it's not a given. It's not like a ladder with all the rungs evenly spaced. Rather it's a substance, a material we try to manipulate just as much as we do the tilth and the fertility of the soil. How many tomatoes can we harvest before the lightning storm arrives? How many can we sell before they rot? How can we get every-body out weeding the carrots this afternoon even though there are all those watermelons to pick? And how can I get November to come more quickly, so that Oona and Wiley and I can take a nap together and the killing frost will give me some hours alone to read?

It's the end of August. We're no longer shaping the rhythm of the season. We merely step into the morning and let the rhythm of the harvest shuffle us along. The products of this rhythm — the song, let's say — are almost unbearably fleeting. The tomatoes get sold, the goat barn gets dirty again, the turkeys eat and eat and eat until they themselves are eaten. The products of an entire season's labor are devoured by shareholders and crew and customers and neighbors and friends, and the residue is harrowed back into the ground.

This is why I like the marks on the John Deere. They're a reminder, a register of all the countless repetitions we perform. The other farmer (a potato grower, I've heard, who recently passed away) and I sitting in the same seat, looking back as we travel forward, resting our palms on the fender to stabilize our bodies, putting sweat to metal, making food for people's bellies — sure — but here's what's left: a small, corroded imprint of our hands. ✍

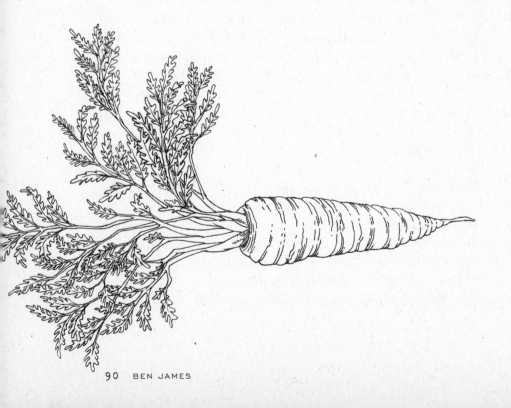

How I Learned to Stop Worrying and Love the Bugs

BY MK WYLE

After three years of "practice" farming and seasonal nomadism, Mary Kathryn Wyle and her husband, Andrew, are excited to be signing their first lease in 2011. They're starting a vegetable and pastured livestock farm near Frederick, Maryland, and even though they're leasing, Mary Kathryn looks forward to planting perennials.

The year 2010 marked a milestone: I had apprenticed for two years, in Georgia and in New England, and had begun to feel ready for farming on my own. Still not prepared for the full commitment ("buying the farm" conveys finality, whether in life or in death), I settled on an organic, grass-based dairy and meat farm in central Massachusetts. The owners were interested in adding vegetables to their offerings, though they lacked the time to tend row crops themselves.

They proposed a deal to my partner and me: his labor with their livestock as barter for my use of their land, equipment, and good name. It seemed an excellent deal, and any doubts I had harbored were quickly erased when I received the results from a soil test. After a childhood of red Georgia clay, the rich, well-drained loam of a Massachusetts dairy farm looked like paradise.

I was growing on "new land," fields that had been pasture for as far back as anyone could remember. I had always envisioned my first year on new land as a sort of horticultural sneak attack, a freebie year in which unsuspecting vegetable pests would miss my fields, like plagues passing by the Israelites. Growing on new land had become, in my mind, talismanic, all but guaranteeing pest-free, low-weed crops. Don't blame the farmers who mentored me for this error — in retrospect, I can only conclude that my delusion sprang forth fully formed from my own overactive imagination.

There is, of course, a kernel of truth in my fantasy — pest pressure is probably lower in your first year, than, say, your tenth. What I neglected to account for in my optimism, however, was that although the bugs didn't know to expect me, I didn't know to expect the bugs.

I knew I had abundant earthworms, sure. I knew I had preying mantises, and beetles, and a local turkey population apparently unharmed by hunting season. But I didn't know — and this was critical — what pests were inherent to my soil. The trick of this is that the first (and decidedly most effective) line of defense on an organic farm lies in prevention, and it's a tricky business to prevent something you don't know is coming.

I felt as though the entire insect world was marshaling its forces against me, competing for every bite I could produce.

One weekend in early May, I ran into a friend who runs a vegetable farm in eastern Massachusetts. She described to me her efforts to save her onion crop, which had recently been beset by onion maggots. She explained to me how the maggots attack alliums by burrowing inside their bulbs and neck. My friend described the tell-tale wilt of her transplants and her horror, upon digging one up and investigating further, to find a sickly white maggot feasting within. The maggots move from plant to plant down a row, she said, and the only sure solution is prevention: covering a bed with row cover fabric immediately after planting.

The next Monday, at my farm, I inspected the onions closely. I noticed some wilt. Digging up a plant, I found an empty husk at the base. I dug up three more. Finally, I found the writhing white assassin within, in flagrante delicto. I went on a rampage. I crawled my way down the rows, digging up every onion showing the slightest sign of wilt. I squished more maggots than I care to think about. The remnants of my surplus transplants and the few remaining onion sets I planted in a fifty-foot emergency bed at the far end of the field. I hurriedly swaddled it in a row cover, looking over my shoulder as though I might catch sight of more maggots mustering.

In June I walked smugly among the potatoes. They were vividly green, lush, and unblemished by potato beetles or leafhoppers. In July we began to dig our potatoes. They seemed fine at first, albeit small from drought. Gradually, however, I began to notice the pockmarks and tunnels that scarred their surfaces. Some seemed almost eaten away, and occasionally an orange, segmented worm ducked into its hole when I held up a potato for scrutiny.

Ah yes, my cofarmers declared, we had wireworms, a common problem when pasture is transitioned into crops.

I, of course, having never before worked on fresh ground, had neither seen nor heard of wireworms. Not that it would have mattered much, my friends informed me, as only time and continued cultivation would alleviate the problem. We dug and sorted our small, wormy potatoes by hand, cursing the wireworms all the way.

June, July, and August all poured on the heat. The tomatoes flourished and the squash and melons grew like magical beanstalks. My fall brassicas — broccoli, cabbage, and kale — bore the brunt of the brutal temperatures. Hidden as they were beneath row cover, I irrigated them less often than I ought to have. To make matters worse, I didn't secure the row covers tight against the ground. On blustery days I noticed the occasional hole in the fabric, edges that ballooned in a wind, and delicate white moths that surfed the breeze. Caught up in the weight of summer's harvest, I ignored these warnings.

In August, I noticed fluttering beneath the fabric at the end of a row. When I pulled back the cover, a swarm of cabbage moths wafted toward the sun. Hundreds of dead moths lay on the ground beneath the brassicas, having beaten themselves to death against the fabric. And my plants? Holey, and wholly infested with the caterpillar predecessors to the moths. They nibbled the broccoli and crawled into the heads; they tunneled through the outer cabbage leaves; they nestled in the creases of the curly kale. After vigorous dunking to shake off interlopers and intense examination of each floret and leaf, I deemed some of the crop passable.

In midsummer, flipping through books on pest identification, prevention, and cure, I despaired. I felt as though the entire insect world was marshaling its forces against me, competing for every bite I could produce. I had nightmares of crop failures, disease, clouds of locusts blotting out the sun. Amid my tribulations, real and imagined, the miracle of that season was the miracle of every year on every farm: plants wanting to grow.

The onion maggots? Gone as quickly as they had come (squished to death, I like to think). The remaining onions filled out round and plump and cured up nicely. With the advent of the cool fall weather and rain, the brassicas returned as if from the grave. They gave me crop after crop of bug-free broccoli, mountains of crisp kale, and enough cabbages to stink up a root cellar far larger than ours. The potatoes, alas, were a sorry crop: small or damaged or both. But we had planted such an inordinately large quantity of them that even after screening out the damaged tubers, we were never short on (small) potatoes.

I concluded the season in a far more stable state of mind than I began it. I try to remember that I can't agonize over any single day of disaster, or a week of disasters, or even a bad year. I'm not a sorceress, responsible alone for the ripening of fruits and the swelling of roots. I'm feeding the soil of my farm — not the plants or the pests. I'm one player among many. In the long run, there are no free lunches on my land — not for bugs, not for crops, and certainly not for me. ✍

The Secret Life of Fruit

BY JOSH MORGENTHAU

Fishkill Farms, the apple orchard started by his grandfather in 1913 in East Fishkill, New York, is where Josh Morgenthau grows apples, peaches, cherries, berries, and vegetables and raises some livestock. Josh studied art at Yale University, and in addition to producing good food for local customers, he enjoys painting — when he has the time.

One of my favorite childhood memories is opening my eyes as the morning sun filtered into the small bedroom of my family's weekend home, a dark brown trailer. It lay far away from our apartment in New York City, among the apple trees on our two-hundred-and-seventy-acre farm, a working pick-your-own orchard in the state's Hudson Valley. Life there was different from in the city, where I went to school and my parents worked. Fall abounded with the musky smell of fruit-filled cold storage rooms and the shouts of customers invading our orchard to fill their bags. Early spring meant pruning time and the rough scrape of apple bark on my skin as I scrambled up the trees to prune the tops. Spring brought the hum of birds, insects, and bees returning to do their work amid an ocean of white apple blossoms.

My grandfather, who had bought the land, grew apples there before embarking on a life of public service. Since then, most of the family in each successive generation showed less and less interest in the land. My father was the only one of the three children to keep his part going, who wanted the farm to continue. His brother had sold his land for development and his sister had let her orchard get overgrown and become a pest-breeding ground for ours.

Although he continued to operate the farm, my father's real career was elsewhere. When our elderly farm manager finally retired, my dad leased the orchard operation to other growers. Over the ten years that followed, it was neglected. The trees went unpruned and became disease ridden. The small leaks in the barn roof grew into holes. And over the years, as West Coast apples flooded the market, my dad was forced to sell parts of the farm to keep the business going. Most other farms in the area closed down altogether. Since the time when he was a kid in the 1930s, my father's orchard had gone from one of a dozen to one of only two in the area.

It was in the context of such losses, and seeing pieces of the farm that were once special to me turned into tract housing, that I wound up, after a detour in the world of higher education, back at the farm. I had no real experience or qualifications other than a head full of ideas and excitement. As my friends moved on to pursue dreams in the Big Apple, I dreamed of growing apples. It seemed like a conjuring act that had been woven into my family history for almost a hundred years, a fundamental mystery to me.

My father and I decided we'd stop renting out the farm and attempt to run the orchard, the complex farm market, and the pick-your-own business ourselves. We hired a farm consultant, a store manager, and vegetable growers. Suddenly I had forty acres of fruit trees on my hands and an incredibly steep learning curve ahead. I added three new numbers to my speed dial — that of our retired farm manager, who was still willing to share his wisdom; that of a friendly apple grower, who at my begging, took me on as his protégé; and that of the Cornell Extension agent who specialized in fruit.

Balance did not come easy. We squeaked by, that first year: We managed to sell a good amount of our crop, despite a series of personnel mishaps and false starts. The old equipment we thought we could rely on spent more time in the shop than in the field. The farm seemed to be in chaos. But by the second season, we had a good team of employees and were heading in the right direction.

Then a fire destroyed our historic barn. It took with it all our storage space and the arsenal of equipment it held. The season that followed was one of the wettest in history. With a pick-your-own operation, rainy days can be even more devastating than crop failure. I realized then that no matter how beautiful your crop is, it isn't worth anything if you can't sell it.

By this point, I was putting everything I had into the farm. I spent almost every waking hour growing or marketing fruit. In some ways,

> Work, risk, and often multiple failures, I found out, are behind fruit's effortless facade.

it was liberating. There was a freedom in bondage; I had all my work cut out for me, nothing else to think about, no time even to question or consider my next move.

Finally, by our third season, things were relatively chaos-free. Still, it was one of the driest in years. This was great for disease control and for our pick-your-own market, but it was bad for a farm with no substantial irrigation system. That year, we ran out of apples — a happy problem. It would have been happier, though, if our crop hadn't been two thirds its normal size.

As I took on the responsibilities of the farm, I became inducted into the secret life of fruit. More and more, I marveled at it: Fruit is nature's purest and most immediate enjoyment, requiring nothing more than a rinse or simple rub on your shirt to clean it. From the fruit-eater's point of view, it's effortless pleasure. It demands none of the slicing, chopping, soaking, or parboiling needed by vegetables. Even on a chemical level, its energy is more accessible, more mobile, with no complex starches to break down. But the immediacy and the sweetness are deceptive. Work, risk, and often multiple failures, I found out, are behind fruit's effortless facade.

I wanted to grow the fruit organically. But beautiful organic fruit is not a gift from nature. If it succeeds at all, it's as a drop of goodness squeezed from a very unforgiving rock: From the moment of petal fall, when the flower's ovaries swell into clusters of tiny fruitlets, apples are as vulnerable and frail as lambs in a pack of wolves. This window of fragility lasts for at least a hundred and twenty days, during which an apple contends with a murderer's row of insects and diseases.

I became acquainted with the despicable apple maggot fly, which with a prick implants its offspring, all gleefully tunneling their way through the fruit in spirals. Then there's the perverse codling moth, which enters and chews the fruit from the butt end. Finally, the dreaded plum curculio, the Houdini of insects: It doesn't disappear, but it lays eggs in your crop and causes the apples to drop and disappear. This is to not to mention fungal diseases, which are even more devastating to your crop. In bad years, apple scab, the major culprit, will defoliate an entire tree.

I wanted to grow fruit without spraying, but it wouldn't be easy. Did I want to lose my whole crop? How great a benefit to humanity would it be if the farm went out of business? Even from an environmental point of view, running tractors, fertilizing with organic fertilizer, and putting untold other resources, human and otherwise, into growing an organic crop, only to lose it on principle . . . well, that just didn't seem reasonable. I quickly understood the root of the disdain many older farmers have had for the organic movement. In growing fruit, reality can quickly crowd out ideals.

For me, though, growing apples was not so much an experience of losing ideals as it was of re-centering them. I set out to grow fruit as organically as possible, but I settled for low-spray integrated pest management (IPM), in which synthetics are used minimally but are not out of the picture. It took two seasons of spraying with conventional materials before the previously neglected trees were healthy enough to start a true organic regimen in one block of the orchard.

In our supermarket culture, fruit has become so visual, so linked to beauty and perfection, that people ignore the fundamental paradox of modern fruit production — high levels of chemicals are the cost of unscathed, "perfect-looking" fruit. In pursuit of this ideal, we've lost a sense of what good fruit might actually look like, cosmetic imperfections and all. I found that many heirloom varieties have some innate disease resistance, which made them a no-brainer for our orchard. Being interested in growing historic varieties and growing fruit organically, I planted thousands of them.

Orchards and fruit trees have a special potential to span great numbers of years and to link generations. It's said that America's longest-lived apple tree was planted in 1647 by Peter Stuyvesant in his Manhattan orchard and was still bearing fruit when a derailed train struck it in 1866. The fruit trees my grandfather planted were pulled out and replaced long ago. But the varieties I'm planting now, against the current of increased yield and aesthetic improvement, are some of the same varieties he once raised. He probably stopped growing them because they were no longer commercially viable — because of their funky flavor, low yield, strange shapes and colors, many of the same reasons they're becoming popular again.

But for all the reverence I feel, I'm careful not to idealize the farming of long ago. Lead arsenate, a double whammy of human poisons, was the number one weapon in the commercial fruit grower's arsenal from the 1890s through the mid-1900s. And prior to that, in the early 1800s, apples looked nothing like they do today. People had never seen anything resembling the apples of today. And getting today's customers to accept apples that bear more physical resemblance to potatoes than to fruit turns out to be even more challenging than is growing them organically in the first place.

New disease-resistant varieties, the product of years of concentrated university research, hold out some of the best hope for growing marketable organic fruit in New York. So, go ahead, plant apples for your grandchildren — just choose a disease-resistant variety if you're not into spraying every seven days. We live, after all, in another world, with different resources, new information, and changing popular tastes. We cannot simply "go back" to the farming of our grandparents. "Remember your roots," my dad always told me. "Don't forget where you came from."

When I first started going to the greenmarkets, he remarked, "You know, Josh, it's amazing. Our family comes from a great line of peddlers and now you're selling your goods on the streets too. It's really amazing that you're continuing the tradition." Taken aback, at first I thought he was pointing out an embarrassing form of social regression. It reminded me of a refrain I often got from people I met: "Is this what you thought you'd be doing after college?" And what university-educated farmer hasn't gotten that question at some point? Usually, it comes with implicit sarcasm, pity, or both. But as I processed my father's remark, I came to realize he meant exactly what he said. ✍

CHAPTER FOUR

PURPOSE

This chapter is about why we farm.

Farming is hard work, good work, but not to be taken lightly. It takes motivation. If we weren't motivated, we'd be doing something easier, more accessible, more acceptable to the rest of society. But here we are, farming — energetically farming, passionately farming. Why? We each have our reasons. It's a process that starts with that first season spent out of doors, covered in bug bites, doing someone else's chores.

Apprenticeship is the portal into farming and, for many, a time of profound self-realization. For one thing, it's often a lot of time spent alone in a field, in a quiet barn in the morning, bent in strain against a heavy load you're not quite up for. It's challenging and it's solitary, with plenty of contemplative space. That space can be difficult; the quiet can be lonely. But that's part of the deal — figuring out who you are when you're alone, what you like, when you giggle to yourself, how you keep your mind still while forcing your body to work harder than ever before, how to comfort yourself, how to train your thoughts along a constructive trajectory — snipping off the side shoots and moving forward. You use the time and the space and the silence to think things through, and to come to some big conclusions. Our education hasn't always trained us for this — but in fact this is the critical self-reflection of being human. Transformation is a big word, but so is what happens on a farm over the course of a season in the inner world of a new farmer.

Ultimately, we come to our own conclusions about life, and about how we want to live it. What role in society feels right and good, which physical space we will inhabit, what it means to farm. For me, it's about having a dependably sensual daily life. I've become a sucker for sights, smells, textures, and rhythms. The sheer materiality of it — greening up pastures, strong round eggs, shiny clean jars and the sounds of their lids, the firm little legs of piglets, the snapping succulent stems of spinach. The swing of it all. I got myself some nice old tin buckets with wooden handles just so I could swing the pig slop better.

It's true — if we wanted to do something easy, we wouldn't have chosen farming. And although that decision to farm might marginalize us geographically (far from the big-city lights) and economically (yoked to our partners, to our animal chores, to our land and the neighbors it comes with, tied to the seasons, the toil, the weeding, the irrigation), it is also freedom. And that freedom, to think for ourselves, quietly, out of doors: That is a freedom we cherish above all.

— Severine von Tscharner Fleming

Purple Flats

BY NEYSA KING

Neysa King works on an incubator farm in Austin with her husband, Travis. You can read more about her, Travis, and Round Table Farm at her blog, Dissertation to Dirt (see page 251).

Friday afternoon at four, I was on my way to pick up my husband, Travis, from Green Gate Farms. At six, we were supposed to be at the University of Texas for a reception. I had been invited as a Normandy Scholar alumna, a history program I had participated in during my junior year. That Friday I had come home from work, showered, and dressed before I left for the farm, as I knew it would be a quick turnaround from getting home at five to being at UT at six. Travis couldn't exactly go in his work clothes.

I don't make a habit of reliving my college days, but this particular program is important to me. The professors I studied under were more than just mentors; at the time, they were who I wanted to be. They taught me to write, and to think about history differently, and they had sparked my interest in human rights and genocide studies, which I pursued in graduate school in Boston. I wanted to show my support for a program that was pivotal in my academic life, and that could be equally valuable for other students. I also wanted to catch up with the professors for whom I felt so much affection. I hadn't seen most of them since I graduated. Some of them knew I had gone off to a PhD program. None of them knew about my current foray into farming, and I couldn't help but wonder how they would react to my career choice.

As I took the MLK exit from Route 183, memories from school were swirling around in my mind. I realized that it had been almost two years since I left school for farming, and I'm still figuring out how to cope with the looks I get from people who were once my peers.

That didn't sit well. Was I buying into all the stereotypes I was fighting against? I wasn't sure how to get past it. What I didn't know as I pulled up to Green Gate was that I was about to find out.

I got out of my car and walked over to our field to check on the carrots. Between the poor germination and the weeds, the picture wasn't particularly promising. As I was thinking of all the weeding we'd have to do on Saturday, Travis appeared with Skip, Green Gate's owner. They walked the rows together, discussing our crop. I had on some pretty purple flats and didn't want to muddy them, so I stayed at the front. When they came back, Skip suggested we re-till and reseed a bed or two. He said the germination was too spotty to justify hand-weeding five beds. Plus, the rains we'd gotten that week would provide a much better seedbed than the dry earth we had been working with before. We should take advantage of a potentially rainy weekend by getting the seeds in now.

```
A part of me was upset that
I couldn't just say, without
equivocation, "I'm a farmer.
I grow your food."
```

It was already approaching five o' clock. To get the beds ready to reseed, we'd have to unhook all the drip tape, till with the tractor, then rehook the drip tape. We'd be late to the UT reception.

"Well, you'd better get used to it," Skip teased. "The weather is going to dictate your schedule for the rest of your lives."

I began walking down the rows now, mud clinging to the bottom of the purple shoes I had bought in New York. Skip was right. The soil was beautiful. A new litter of carrots might come a few weeks late, but they'd be more uniform and healthier, and they'd take less time to weed. I didn't want to miss seeing my professors, but I couldn't leave my field. I kicked off my shoes and rolled up my jeans, and Travis and

I began taking off the drip tape. Once Travis got the tractor, I became so excited when I saw dark, heavy soil fluffing up behind the tiller — so unlike the sand blowing around a few weeks ago — that I wanted to redo the tilling of three of the five beds of carrots. We'd keep the two beds with the best germination. We arrived home just at six o'clock. My feet were covered in dirt. I was happy.

We did make it to the reception, albeit late. I had thrown on a black jacket and skinny jeans, but I still had dirt under my fingernails and my hair was windblown. Just as expected, I met the new class of Normandy scholars, and began mingling with my old professors. But when I got the inevitable question "So what are you doing now?" I felt the need to tell them I had been in a PhD program, and that I had in fact received my master's degree, before going into farming. I got different degrees of support and incredulity. A part of me was upset that I couldn't just say, without equivocation, "I'm a farmer. I grow your food."

At the same time, the night was freeing for me. I hadn't realized it until that point, but I'd been idealizing my old professors. To see them again, eating hors d'oeuvres, discussing academic politics, and fussing over college kids that looked so young to me now, I remembered graduate school more vividly than ever. And I remembered that I had, with a clear mind, decided to leave. My professors, I could finally say, weren't doing anything more or less valuable than I was. They were just doing their jobs. And back at Green Gate that afternoon, I, for maybe the first time without looking over my shoulder, had done mine. ✿

Write It Down

BY JENNA WOGINRICH

The fiddling shepherd of Cold Antler Farm, in Jackson, New York, Jenna Woginrich is an office worker by day and a farmer by passion. On her six-and-a-half-acre homestead, she raises sheep, chickens, geese, rabbits, vegetables, and bees. The farm's main business is the fiber CSA that produces yarn for shareholders all across the United States and Canada. She's also the author of several books on country living for beginners.

There's a fire in my woodstove, and between that and two glasses of homebrew, I'm very warm tonight. I just ate a simple dinner of pasta and tomato sauce, then (with a slight buzz and a full belly) I pulled my fiddle off the shelf and played a few Irish tunes to light up the room. In a little white farm house in Jackson, New York, "The Scartaglen Slide" and "Man of the House" trotted out with my bow while the dogs wrestled on the kitchen floor. I did a little dance and hopped about with them as I fiddled, the fray of gnashing teeth and my laughter tearing up the peace. It was quite a sight.

What a life this has become! Over my pigtails, I'm wearing a warm hat that I made from wool off the sheep in my pasture. There are eggs in my fridge from the birds in the coop, and there are chickens (and a rabbit) I harvested in the freezer. Besides the meat I've raised, I've made bread, sauce, jams, cheese, beer, cider, and pies. There is honey I pulled from my own hive, and a truck in the driveway. I have a fine pair of geese. I even held one of their just-born goslings in my palm this time last fall. I've grown a garden full of vegetables and held pumpkins as big as bobcats. I've hunted pheasants and shot at foxes. I've heard coyotes sing in the pale moonlight and watched them from the edge of a sheep pen with a crook and a lantern. I caught a native trout on a dry fly and I know when a river is angry. I've raised rabbits. I've written books. I've sewn clothes. I've ridden a dogsled in the blue glow of a winter sunset, and I know how it feels to bottle-feed a baby

goat on a porch during a spring rainstorm. I can now sit high in a
dressage saddle and do a posting trot with a sixteen-hand horse.
A little black-and-white rocket of a dog runs about as I write, and he's
the future of this farm: my business partner, Gibson the Border collie.

We have a CSA in the works, we shepherds, and soon we'll be
sending out packages with wool and thank-you letters to our inaugu-
ral subscribers. There are sheep on the way, too. Those ewes will
be heavy with lambs and I'll bring them into the world this spring.

Tonight my plans don't involve any hot dates — and certainly
nothing like a night out on the town — this is a night on my farm. Cold
Antler Farm. This place didn't exist in a gasp five years ago, and tonight
I'll be reading about the proper bedding and pen setup for a pig.

When I knew a farm was something
I wanted, I sat down and wrote out
exactly what I hoped it would be.

Tomorrow, I add a little swine to the mix. It seems as normal now
as deciding which fabric softener to buy from the grocery store. This
is my everyday life.

I've been told that I'm a goddamned fool. I must be. Only a fool
would be living like this, doing all this, and dancing with dogs to tunes
no one else knows anymore. You can call me whatever you please.
I'm not changing a thing about this messy life. I like messy. It suits me.

Listen, I don't have much money, and I'm nobody's Daisy . . . but
I'll be damned if I'm not happy tonight. I feel like the wealthiest beast
in the world. And you know why all this happened? It happened for
two simple reasons, and I believe this with all my heart. I landed here
because:

1. I always believed I would (not could, not might, but would).
2. And because I wrote it all down.

Something that stuck with me in college was a blip I heard on the radio one night. A person was telling someone on NPR that if you want something to happen with your life, you need to get out a pen and paper and write it down. He said that only 2 percent of people with goals actually take the time to write them down, but out of that 2 percent studied, 90 percent achieved their dream. Something about the certainty of pledging it to yourself made it more real to the people he observed. I wanted to be in the 90 percent of that 2 percent.

When I knew a farm was something I wanted, then, I sat down and wrote out exactly what I hoped it would be. I wrote about a hillside outside my window, about the sheep, about the black-and-white dog by my side. I drew a pickup truck parked outside, and a veggie garden alive with a lush bounty.

Okay, so not everything came true, but the point is that most of it did. I carried that piece of paper with me until it naturally disintegrated into scraps. It was my totem, my prayer. And I think because I physically held it on my person, I could never forget it was there, and always being on my mind forced me to always strive toward it.

That said, it's not a magic trick. It wasn't exactly as if it fell into my lap. Nothing was given to me; I had to earn it. I had to wheel and deal, and beg, borrow, and steal to make it happen. But it did. I pulled it off, paycheck to paycheck, a little at a time, until it rolled into something so epic it wore me down and built me up again. Somehow, I got a mortgage, a collie, a truck, some land. Somehow, I raised a barn. There are fences outside and a CSA on the books. Thanks to the help of many hands, my amazing parents and siblings, friends, blog readers, thoughts, prayers, and (I think) my daily diary online, my aspirations went from a pipe dream to a steam engine. If it was something a girl from Palmerton, Pennsylvania, could get, you can too. I promise.

So if you are someone who wants your own land, your own farm, I urge you to sit down and write what you want, tonight. Write it all down, fold it up, and put it in your pocket. It might take five years before you're in your own kitchen dancing with a Border collie. But hell, those five years are coming, one way or the other. Might as well have a farm at the end of it all.

And keep dancing in your kitchen. It can only help. ✍

Growing Not for Market

BY DOUGLASS DeCANDIA

In the Lower Hudson Valley of New York, Douglass DeCandia lives, grows produce, and raises animals. He manages a farm operation and oversees a food-growing program with the Food Bank for Westchester to provide fresh produce to individuals with limited access to good food and to support a hands-on education for young adults and students.

I remember the day, the conversation, and what was being done when I was asked a really good question.

A neighbor and I were in my field, among the growing leaves, young fruit, and earthen smells of early summer. I was tired and my will was fading away from the harvest at hand. The Swiss chard I held, as wholesome as it was, was heavy. There was a weight put on everything I grew: It all had to be and look a certain way in order for it to go to market and sell. Producing for high quality wasn't what was bothering me; it was the fact that most of the people I was selling to were more concerned with what the produce looked like, tasted like, and how it was grown than with how much nutrition they were getting from it. The weight from that made me feel as if I was doing something I didn't really want to be doing, even though I really did want to be farming. Maybe I just needed to do it a different way.

"Why don't you just donate the produce?" a friend asked.

It was a simple question, but it ran deep for me. It made me realize that not only did I need and cherish the physical work of farming, the mental exercise of running an operation, and the spiritual connection with land, but I also wanted to share all this with others who had less access to good food.

"Why don't I do this?" I asked myself.

"Well, because I can't make money donating food," a small voice in me responded. "I want to farm full time, and if I can't make money doing something full time, I can't do it full time. It can't happen."

That didn't feel right. I knew it could happen.

We all need good food — to nourish our mind, spirit, and body so that our whole being can function at its fullest. Malnutrition doesn't discriminate. It plagues the rich as much as it does the poor; young and old suffer from it. Although many of us have access to good food, way too many do not: the homeless, the young, the hungry, the poor, the elderly, the handicapped, the most vulnerable. Without proper nourishment, the ability to think and make good decisions, to maintain and build physical strength, to develop the will and spirit is greatly weakened. With good food comes empowerment.

My ambition became to grow nourishing food with and for those with limited access, while also meeting my own needs. I pursued private and public funding, regional food-justice organizations, and friends and family, trying to find the land and money needed to begin a farm. I developed relationships with good people and organizations, found a place on committees and boards, and was eventually offered a position with my county's food bank.

The Food Bank for Westchester (New York), with its mission to lead, engage, and educate the county in creating a hunger-free environment, wanted to increase the quantity of locally grown produce it provided to its clients. It hired me, gave me a livable wage, and offered me the opportunity to farm and *not* grow for market. ❧

What to Do If You Think You're Not Good at Anything

BY A. M. THOMAS

A first-generation farmer and writer, A. M. Thomas comanages East Hill CSA, a
small-scale vegetable, fruit, and bread operation in rural Middlesex and Rochester,
New York. He maintains what he describes as "a serious literary blog" called Wear
a Wax Dustcoat (see page 251).

Sometimes people ask me why I farm. I tell them different
things. To some I say that, biologically, we are meant to be farmers.
"We've been farming for thousands of years. Why stop now?" I say.

To others (seeing an opportunity to shorten or end the conversation as quickly as possible), I say that I farm because I like good food.
"Can't argue with that," they say, thankfully.

To a third group of people, usually those most interested in
farming, I explain that when I was younger I made a list of jobs I could
imagine myself enjoying. I tell them the list included "small-scale
organic vegetable farmer" and that I somehow fell into it. I add some
esoteric, overly idiosyncratic items to my fictional list of self-actualizing
professions in order to make them laugh or to distract them. I say
that besides farmer, on my list were rapper, astronaut, lonely graduate
student, writer, playwright, lonely history professor, and lonely Civil
War reenactor. I explain this maniacally, with eyes wide, until whoever
asked the question starts talking about himself or loses interest.

To the fourth group — those with whom I'm most honest — I
shrug and sadly mumble something about not knowing what else to
do. "I could probably be a good janitor, maybe," I say, almost inau-
dibly, "but I don't know what else I'd be doing. I'm not really good at
anything."

I grew up in somewhat urban New Jersey, about twenty miles outside of Manhattan, and didn't have a lot of interaction with nature. My dad kept a small vegetable garden in my aunt's backyard until I was nine or ten and then he stopped. I remember helping him in the garden a few times and liking it.

I ate a lot of processed food. I liked Toaster Strudels and Pop-Tarts. I liked bread. I put ketchup on most things. Most of the time I felt really awful. I wondered why my stomach hurt so much. In high school I went to a digestive specialist, who gave me a cup of high-fructose corn syrup to drink. I got sick almost immediately. He told me I had an HFCS allergy and "probably irritable bowel syndrome or Crohn's disease" or something. It seemed that most of the food I was encouraged to eat was poison to my body. I was frustrated by my stomach and, though I didn't realize it then, by the food system I was trapped in.

The best things in life — growing your own food, living and working with your neighbors, being outside in an open space — are being lost.

Being sick showed me that there's a lot wrong with the way things are set up and maybe, I thought, if we do things differently, there's a chance we could get it right. I discovered subculture. I learned that there are alternative ways to eat, which, it turns out, is how most people in history have eaten. Sometimes I wished I'd been born a hundred years earlier.

After college, I left New Jersey to become a farmer. Through WWOOF (Worldwide Opportunities on Organic Farms), I discovered a farm about six hours northwest in the Finger Lakes region of New York. The farm, where I still live and work, is called East Hill Farm. It's a project of the Rochester Folk Art Guild, an intentional community of craftspeople and farmers who have lived together in Middlesex, New York, since 1967.

So, I made the odd, difficult transition from a life rooted in urban culture in New Jersey to a rural, agricultural lifestyle in an established intentional community. It's a transition that I'm still trying to figure out. I've learned more practical skills than I ever thought I would: bread baking, logging, vegetable and fruit production, woodworking, operating a tractor, canning and food preservation, beekeeping, raising and slaughtering pigs, raising and slaughtering chickens. I've learned how to live by myself in a one-room, "off-the-grid" shed through the winter. I've experienced love and heartbreak and made great friends. I've been more confused than ever before. I've discovered that I have much to learn about human interaction and relationships.

I'm now on the verge of my third season of farming. It's the best job I've ever had, though also one of the most puzzling. Sometimes farming feels simple — like the crops grow themselves, and it's almost a gift that this work exists for us. I've thinned beets while lying on my side in beautiful June weather and thought, "Farming can be lazy and relaxing, I guess." Other times, farming seems impossible. It feels like there is so much that has to go right — too much — for it ever to work. But despite my inexperience and lack of knowledge and small stature and self-deprecation, so far I've somehow made it work.

If you sometimes feel that you're not good at anything, consider becoming a farmer. (It's probably what you're supposed to be doing, anyway.) You'll discover that you're actually good at many things. You'll learn many skills that make you feel fulfilled and proud of yourself and then you'll realize that these are all the skills that are being forgotten. The best things in life — growing your own food, living and working with your neighbors, being outside in an open space — are being lost.

Know, also, that farming is tough. Some days, maybe most days, you'll feel overwhelmed. When your crop of onions is failing and your tomatoes have blight and the weed pressure on your winter squash is mounting and you can't stand the people you work with (or, worse, the people you work with can't stand you) and your livelihood depends on this food, you'll feel overwhelmed and even afraid. But you'll also feel a fullness. Your life will feel different from how it would if you were a young person living in a city, working in an office, going to bars and restaurants. You'll know what quiet is and you'll be able to go outside at night and see darkness. Your body, at first weak from the winter or the suburbs, will reject your work. Then, after struggling, it will embrace it. You'll eat good food. Eventually, you'll ask: "How do I live well?" And we need you to answer that question. We desperately need you to. ✑

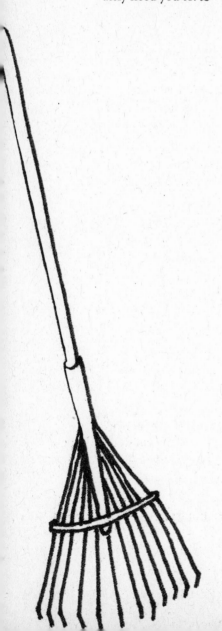

Farming in the Web of Interconnectedness

BY SARAJANE SNYDER

Although she is currently the farm manager at Green Gulch Farm in Muir Beach
California, Sarajane Snyder considers central Pennsylvania to be her home.
She recently found another list (in addition to the one on page 116): What Does
the World Need? Listening, Slow, Less, Breath, Sharing. Free farm stands.
More art space. Less consumer space. Reskilling.

Farmers everywhere:
You are creatures within creation
Do not be afraid to consecrate your farmland
If you have heard the quiet sacred whisper
If you have witnessed the radiating energy
If you have felt the happy exhaustion of
Taking the long way home via the land
Do not be afraid to give a little amen
For what is all around us
Sustain us

Sometimes we hide treasure for ourselves, forget it completely, and
then delight in rediscovering it. On a blowing winter night, I read
through last season's farm journal to prepare for the coming season.
Amid notes about seeds to try, calculations on pounds of potatoes to
order, quotes from books I was reading, notes on planting practices
and pick lists, updates on our organic systems plan, reminders on
tractor maintenance, I find this all by itself on a page:

a good list
human health
community engagement
cultural knowledge
ecological restoration
political empowerment
economic stability
spiritual enlivening
(dependent on networks of relationships & connections)

This is indeed a good list. It's a broad list. An idealist's list. This is a list that lays out the bare bones of why I farm and why I hope every person who has a hair of interest in farming can be supported and encouraged to jump in and explore the many paths that crisscross through the field. I have come to believe that through a lot of rejuvenating and a little reinventing, the art of farming in this country can immediately ameliorate some of our more persistent and idiosyncratic problems (poor health, isolation and loneliness, cultural dead zones, ecological degradation, political hopelessness, economic turmoil, and spiritual impoverishment, to name a few). Many people have written and will continue to write about (and enact!) the role of small farms in creating antidotes to these problems, but I think that the antidote of spiritual enlivenment is perhaps the least often addressed. So I thought I'd give it a few words.

Although I grew up in central Pennsylvania, a region with plenty of religion and plenty of farming, I came all the way to California to, in a sense, reconnect with these parts of my heritage. As a young farmer, most of my training has taken place within the context of Zen Buddhism as practiced by a community settled in the San Francisco Bay area for almost forty years. I arrived at Green Gulch Farm, part of the San Francisco Zen Center, in 2005 with no knowledge at all of Zen and only the briefest elbow-brushings with organic farming. I was applying for the farm and garden apprenticeship, and I'm grateful to have been invited to stay on for six months. During that time (and the seasons that were to follow), I got rich: I learned what it means to live in community; I learned about religious forms and ceremonies;

I learned about silence, stillness, devotion, and love. I also learned what a joy it can be to grow food, teach others, care for the whole landscape, and tap in to a very rich heritage of small farmers.

The source of all these riches? Please see the bottom of the good list: "(dependant on networks of relationships and connections)" is written there like an afterthought, but it's the foundation.

For me, one of the most profound teachings of Zen Buddhism, and one that I know will resonate with all ecologically minded farmers, is the awareness of all beings and our interconnectedness. Every single day at Green Gulch we invoke the presence of "all beings," from which we as individual farmers and truck drivers and web designers and kindergarten teachers are not separate. In fact, there is no separation at all. My cat sleeping on the bed, the mites in his ears, the microbes in my composting toilet, the ancient beings whose fossilized bodies fueled my delivery van into the city this morning, you, your parents, all the people you've ever kissed or high-fived or tried to ignore on the bus, giant redwoods, ephemeral mushrooms, molds, grasses, ravens, poison oak — we are all interconnected in this great net of being. Just by simply existing, we support and are supported by this interconnectedness, but I say it becomes all the more powerful when we give and receive this support with awareness.

Food links our bodies and the body of the earth.

What better way to study the complexities and paradoxes of interconnectedness than by studying food? Food is an intricate dance of causes and conditions — human acts, vital seeds, four elements, and a vast array of plant, animal, and microbial helpers and hindrances. And that's only to produce the food. There's also the vast world of harvesting, of selling, of procuring, of preparing, of consuming, of disposing of the wastes of eating — all acts that have been swollen to an international and industrial scale.

Although a globalized food system might seem all the more wondrous an example of our vast interconnectedness, I don't think it serves the project of witnessing and becoming more deeply aware of our own value as participants in the health and vitality of other beings. When the food we eat was produced in a place (whether it be a distant country or the unknown interior of a factory) we cannot conjure up, our food — one of the greatest links between our bodies and the body of the earth — becomes another anonymous product to be taken for granted and removed from any living context.

Here at Green Gulch, I am honored to serve on the farm, which has grown and evolved in tandem with the *sangha* (a Buddhist term for community or congregation). Some of the abbots and former abbots of the Zen Center used to work on the farm, the land has hosted thousands of temple visitors over the last decades, and our farm altar is part of many temple rituals throughout the year, including a seed-sowing ceremony in February and a Thanksgiving Harvest ceremony in November. During the summer months the entire community, instead of going to *zazen* (meditation), comes to the farm at dawn to participate in silent hoeing before breakfast. Some of our apprentices have gone on to become farmers, some have gone on to become priests. It has been vital to my development as a farmer to see not only how enmeshed a farm and a community can be, but also what it looks like when the community holds the land as part of a sacred tradition. This experience, combined with the teaching of interconnectedness, has opened my heart to the reality that all farms are sacred places, not just the ones that "belong" to a religious community.

Farmers, understand what you're doing in the context of interconnectedness, of caring for multitudes of beings. Take refuge in the care you are generating and the sustenance you are providing, for humans and bees and microorganisms, for gophers and fish and spiders. Our dirty work is good work.

Farmers and nonfarmers alike, you have the opportunity every single day to be reminded of the amazing network of beings that nourish you. And to give thanks. There's a bumper sticker I particularly like, despite my current non-theistic status. It reads: ARE YOU WELL FED? THANK GOD . . . THEN THANK A FARMER. We recite a more ecumenical grace at Zen Center: "We reflect on the effort that brought us this food and consider how it comes to us, the work of many people and the transformation of other forms of life."

If you stop, even for a moment, each time you're about to eat something and think, simply, "Someone's hands picked this food," it would be the beginning of a brain pattern of recognition. You don't have to believe in a god, you don't even really have to give thanks. You are just recognizing that you exist in a state of dependence on other beings, and that other beings depend on you as well.

May all beings be equally nourished. ✍

Farming with Two: Pleasure and Independence

BY EMILY OAKLEY AND MIKE APPEL

As the owners of Three Springs Farm, a diversified, certified-organic vegetable farm in Oklahoma, Emily Oakley and Mike Appel cultivate more than fifty crops on five acres. They sell their produce through a farmers' market in Tulsa and a one-hundred-member CSA.

Neither of us grew up on a farm or even had so much as a family garden.

We came to farming with a passion for its intersection between social justice and environmental issues. Internships on large-scale organic farms in California sealed our fate, as we fell in step with the rhythms and way of life farming offered. We resolved to use only our four hands to run our own farm, adapting lessons learned in California by modifying them to a smaller scale.

Even though we own twenty acres in rural Oklahoma, we cultivate only three and a half acres of diverse annual vegetable crops and have two and a half planted to perennials. We take pleasure in the independence of farming. It's our full-time job; we have no off-farm income, and we like it that way. It's our goal to operate as a two-person farm with no hired labor or interns. We have designed our farm, business, and relationship around these principles.

We aspire to be as small as possible while still making a living. We believe that bigger isn't always better, especially in farming. Bigger means more worries and responsibilities, and it doesn't always equal more money. Instead of continued growth and expansion, we strive for efficiency in time and costs. Like any other target, it's a work in progress and something we tweak each season.

Why make a point of farming without interns or hired labor? We like keeping our hands in the soil, not in the office. We don't have to busy ourselves with extra paperwork such as workers' compensation and payroll, and we don't spend time and energy managing people. With just the two of us, things get done the way we like them done. Obviously, internships and apprenticeships are an essential part of growing new farmers, and our hats are off to those farmers who can make the commitment to providing a place of learning and encouragement for young would-be farmers. We appreciate their farms; we just don't want to be one of them.

Not having employees also gives us the ability to stay small. The higher the labor costs, the bigger the farm needs to be to pay for and justify the expense. Lower income needs mean we require less land to farm, and that made purchasing a farm more affordable for us. Also, it's important to us to have a meaningful off-season; we want time to be involved in activism, for friends and family, and to rejuvenate ourselves. Because we don't have to worry about keeping people employed over the winter, we can stretch our off-season to fit our schedules.

Being a two-person farm certainly has its limitations, though. If anything's going to get done, we have to do it. No sick days on our farm! Got a sore throat? Suck on some Ricolas. Without plenty of extra hands to rely on, it's easy to get overextended. There's also an automatic cap on our income — we can earn only what we can earn without relying on help. And at times, it gets lonely out there in the field with just the two of us and the veggies. Farming without the buffer of other people can occasionally be a strain. Emily might be chattering away while Mike is secretly dreaming of peace and quiet. We are together *all* the time. Generally, we like it that way, but sometimes it sure would be nice to have another sounding board.

Any way you do it, farming takes serious talent and commitment, but farming with two has unique requirements that farms with labor don't have to be quite so concerned with. There's no boss, so cooperative decision-making is a must. Talking through choices and coming to consensus are necessary skills.

Efficiency and good time management take on a whole new meaning. Investing in equipment to minimize labor goes a long way toward

With a two-person farm, there are no sick days.

performing tasks quickly. Weed control becomes something of a science, with tractor-mounted cultivators reducing the need to hand-weed. Likewise, fertility management and variety selection enable us to grow fewer rows and achieve better yields. Harvesting, washing, and packing produce are refined each year by calculating the hours required to pick and prep each crop.

Probably the most essential ingredient to a financially practical two-person farm is a strong direct-marketing outlet. One hundred percent of our produce is sold directly to the consumer through farmers' markets and a CSA. You'll never grow enough volume to sell at wholesale prices and still earn a living as a two-person farm. Selling retail is indispensable, so you'd better be good at it.

In addition to the broader philosophical decisions, there are operating choices we make that help keep us viable, both economically and emotionally. The litmus test for everything is always time. Our greenhouse is a small, inexpensive, homespun design. We plant in one-hundred-and-twenty-eight-cell plug flats and never pot up. We use the tractor to do as much as possible, from tillage and bed-making to transplanting and cultivation. Initial equipment investments quickly pay for themselves in saved field hours and less-sore backs. Cover crops and crop rotations are a three-in-one elixir: fertility, weed control, and pest and disease prevention. They're very cost- and labor-effective — all you need is a broadcaster or seed drill and a mower. Although we plant more than fifty varieties, we focus on anchor crops to carry us through each season and generate cash-crop revenue for each market month, going from asparagus (April), to strawberries (May), to blueberries (June), and ending with tomatoes (July and August).

We farm because we love it, and we want to continue loving it. Burnout is our nemesis. At the end of each season, we ask ourselves what we need to do to ensure that we'll continue loving it into the future. One of the first things we did was eliminate our fall CSA share. Although it's lovely weather to be outside, we decided we would rather have the time to work on farm or community projects rather than earning extra income. Securing a long-term land arrangement gave us the opportunity to indulge in perennials and plan a farmscape.

After buying our farm and moving farther from our marketing outlets, we stopped wholesaling to stores and restaurants. Their purchases were too small to justify the time involved in obtaining and delivering orders. Now the chefs who want our food come to the farmers' market like everyone else. Rather than offering a preselected basket, box, or bag of produce for our CSA shares, we transitioned to a system in which our members pay up-front in the winter and then "shop" at our farmers'-market stand throughout the season off their credit balance. This translates into much less stress over crop quantities and timing and more flexibility for our members. We attend the farmers' market twenty-two out of its twenty-six weeks. We find that five months of marketing is just the right amount for our soil and our souls. We dropped our slow weekday market and now focus exclusively on the higher-earning Saturday market.

Over the course of our eight seasons, our strategy for staying small has evolved. We share those plans with our customers through the CSA newsletter. Every decision to cut something out, be it an unusual variety or a particular farmers' market, has its disappointed customers. We've lost a few over the years, but most are upliftingly loyal. They want us to be here for the long haul as much as we do, and they stand behind our choices.

We expect our experiences, feelings, and goals to be perpetually dynamic. We enjoy working together and collaborating on living our dream. Our simple lifestyle reflects both our modest income and our beliefs. We don't need a lot of money to live happily and well. We don't want to sacrifice the farm dream for the bottom line.

Each season someone inevitably approaches us about interning or working with us. We are honored to be asked, but every year we return to the same answer: Farming with two is our foundation. Farming small is not for everyone or every situation, but it's more sensible than it may at first appear. Focusing on a two-person farm enables us to realize the independence and pleasure that agriculture is all about. ✍

CHAPTER FIVE
BEASTS

If our great-grandparents were still around to impart their farming wisdom, one thing is certain: They would insist that a farm without animals is not a farm. Rudolf Steiner, the father of biodynamics, would agree. Cows, chickens, pigs, ducks, horses, mules, goats, turkeys, oxen, sheep: They provide protein — meat, milk, eggs! — fertility, and horsepower. On the farm, they are the cause of our most acute heartbreak and the source of our comic relief.

The cast, of course, is not limited to domesticated vertebrates: remember the industrious bees, the aristocratic praying mantises, and all the other indispensable pollinators and beneficial insects that grace our fields. Remember, also, the coyotes, the tomato hornworms, and, yes, the gophers, that raze our crops, massacre our henhouses, and taunt our sanity.

In spite of all our farmerly efforts to keep things businesslike, animals have a way of insinuating themselves into a personal relationship with us, in a way that plants don't. Resisting anthropomorphism is tricky. And how can you blame us, when that wily, conniving, goddamn gopher is laying siege to our livelihood in the back forty, as the rooster struts his hot stuff around the barnyard, at the same time as the doe-eyed Jerseys are batting their long lashes at us while generously proffering their frothy milk in the parlor? We talk to our animals, we (the gods forbid) name them, and before you know it, we find ourselves embroiled in all kinds of multi-species relationships.

And frankly, that's how we like it.

What it means, though, when we take up animal husbandry, particularly in the context of meat production, is that we've set ourselves up to face a certain truth: that our livelihood is staked on that happy pack of gamboling spring lambs, and, more specifically, on their timely demise. Killing animals — or occasionally, seeing our animals killed — forces us into a cosmic wrestling match with a few of the larger themes in life: birth, death, and the sometimes uncomfortable human power we wield during the interlude. It's one thing to grow okra — to start the seeds, tend them to maturity, and harvest them with a sharp, skillfully wielded knife. It's another thing to raise broilers — to hatch them, tend them to maturity, and harvest them with a sharp, skillfully wielded knife.

Animals, cute as they may be, are not for the faint of heart, and a real day-in-the-life on the farm will cure any agrarian romantic of certain bucolic notions. It takes some resolve to spend a bloody day castrating lambs, docking tails, and tagging ears. Your "ick" tolerance has to be pretty high to trim up a herd of goats with hoof rot, flick maggots out of a festering wound, or clean out a hog pen. And if you don't have it in you to religiously exterminate rats and mice, well then, you might as well turn around and head back for the cubicle.

If these essays tell us anything, it's that our animal interactions are not just one thing all the time: not just cute and entertaining, or gross, or morally complicated; they're all of these things. Animals are sentient beings, like us. They have intelligence, as we sometimes do. They give us protein, honey, storytelling fodder, and, perhaps most important, an opportunity for us to explore our genuine humanity.

— Zoë Bradbury

Reflections of a Rookie Farmer

BY JUSTIN HEILENBACH

For several years, Justin Heilenbach has been farming in Oregon's French Prairie region of the Willamette Valley. In 2010 he started Farm. (that's correct, Farm with a period) with Terra Senter. Justin and Terra have since moved Farm. to Vermont, to pursue their dream of the American Family Farm . . . and to see if moving three thousand miles from Oregon will help with their gopher problem.

The following story is based on entries from my farm logs, notes from my first year of solo farming in Oregon's Willamette Valley. Mine was a venture growing organic vegetables on about a tenth of an acre in the French Prairie region, about twenty-five miles south of Portland. On reviewing the logs, I estimate that 47 percent of my time that first year was occupied with concerns about the western pocket gopher. Whether I was talking about him, watching him, hunting him, lamenting him, or just plotting against him, it was 47 percent.

Actually, this story is only half the truth. The other half can't be told without a laundry list of curse words. If you want the rest of the story, the gritty and heroic part, you'll just have to take up farming and find out for yourself.

The paradox is this: One of the simplest ways for a farmer to reduce a gopher population is not to farm.

METHODS

You can shoot them. My neighbor told me he gets them with his .22. You can trap them. I saw some guys doing that in the field next door. Poison them. I saw a sign on the highway advertising gopher-poisoning services for twenty bucks. You can smoke them out with sulfur and a torch. (Peter, the owner of the farm, swears by that method.) You can bury metal mesh under the field to keep them at bay. How much mesh would that take to protect my field? Would I need it under the whole field or just on the perimeter, sort of like a moat?

My first attempt at eradicating
gophers involved hunting them with
nothing more than a shovel. This
was not the right weapon for the job.

For a while I considered building barn owl boxes in the field, in hopes that the owls would come and do the work under the cover of night. I heard that one barn owl could eat two thousand gophers in a year. Adding more dogs and cats to your farm is another method to employ. My buddy Nate says cats are really good at hunting gophers. Then there's the possibility of flooding the gopher tunnels, *Caddyshack*-style, but where does all that water go? My brother Tim — who, as you may be able to ascertain, is not a farmer — proposed a plan to plow the fields at a depth of six feet because I told him that gophers den at roughly this depth. The only question is, "Which tractor implement to use for this mighty scheme?" Some of my farm friends prefer the strictly antagonistic approach of throwing their cigarette butts down the gopher holes. And a neighbor at the farmers' market told me not to bother farming at all until I was sure all the gophers were dead.

My first attempt at eradicating gophers involved hunting them with nothing more than a shovel. This was not the right weapon for the job.

SEASONAL NOTES FROM MY JOURNALS
Spring

I talked to Noel this morning (Peter's wife and the gracious steward of the farm). She said the gopher ate almost all her garlic and she pulled the rest because she would rather give it to the chickens than let them have it. I told her my garlic was fine. She seems paranoid.

My plot looks good. The beds are shapely, weeds are at a minimum, and most notably the gophers seem to have largely overlooked my location. Am I like a gopher whisperer or something?

Early Summer

I shot a bunch of gophers this week. I told Noel I was pretty sure I got them all.

I didn't get them all.

Summer

Today I harvested potatoes and realized I now might have a little gopher problem. How many could there really be, though?

A lot.

Shot more gophers today. Gophers have terrible eyesight. You can get within a few feet if you're sure not to move much when they surface and look around.

I've graduated to trapping gophers. I'm getting some and chasing out the rest with my bad attitude.

Trapping, I have come to learn, is by far the superior method. You can place the traps in fresh holes, in areas with crops that are particularly vulnerable to a gopher's big weird teeth and appetite. And traps are far, far, far less time-consuming than walking around hunting them. Although shooting the gophers enables the farmer to witness their demise, with their pouches full of garlic, carrots, parsnips, and the like (the western pocket gopher has a large storage pouch on either side of its mouth for efficient harvesting of my vegetables), and this is very satisfying, it can be a slippery slope toward an unhealthy perspective on the matter.

Late Summer

I can't believe gophers will eat the whole root from a chard plant! I mean really, the *whole thing,* like nothing's left but some wilted leaves. They ate half the salsify, too. Did you know they also like parsley root, leeks, carrots, parsnips, and onions? Who likes parsley root? I'm going to plant winter carrots and parsnips anyway. I'm pretty confident I can keep them at bay.

I'm really excited about salsify. Nobody else at the market has it and a couple of chefs have come to my stand and been really impressed that I'm growing it. I can understand the gopher getting some of the leeks and stuff, but I can't help but feel that the salsify is personal (taking this personally by way of projecting intent is part of that slippery slope, to be discussed shortly, right after the gopher finishes eating my parsnips).

Fall
Parsnips: gone. Winter carrots: gone. Chard: gone. Beets: gone. Kale: holding steady. My attitude: suspect.

Winter
If a gopher eats the last of the parsnips in the field but the farmer has long since given up, is it still annoying?

INSIGHTS
Farming is as much a philosophical endeavor as it is a confluence of skill, knowledge, luck, technique, and effort. That said, a balance must be achieved.

Gopher management is necessary, but not to the extent that the farmer assumes a role of preeminence over the land and the larger order of things, in which he is equally as insignificant as the gopher, lest he become the agent of his own demise, or that of his neighbor.

Divergent points of view can lend themselves to friction in forms as varied as my farm plot. Carrot feels that Tomato takes up too much space and won't share the sunny spots. President Potato is bewildered with Minority Leader Leek over the recent redistricting of the new plot. Farmer wages war against Gopher's seemingly intentional efforts to destroy all things that he values.

To the contrary, divergent points of view also carry the inherent potential for fresh perspectives. Farmer seeks opportunity in rich soil with ample amendments for prolific plant growth. Gopher seeks opportunity in rich soil with ample amendments for prolific plant growth. It is at my hand that this little paradise exists; who am I, then, to turn around and curse anything else that might also be positioned to benefit?

While I transplant the kale, sow the lettuce seed, harvest the beans, there are international disputes at play, disputes that echo my own little controversy right here on the farm. Where to draw the line?

How far am I willing to go to achieve an end? Do I want to bring the glorious salsify to market at all costs? What is my role? Where do I fit into the grander order of things?

To be sure, I have an obligation to try to coax my plants toward market, but of equal importance is my responsibility to steward the land, and to accept that my agenda is not so special from the vantage point of another. Somewhere in this discussion there is a dividing line: where once he may have been burdened, now he is the antagonist of all things.

I've decided to strive for some form of balance. I want to be a wise farmer. My girlfriend, Terra, says she won't stay with me if I'm going to be a grumpy old farmer. So I have a choice to make every day. It's not so much about whether to manage the gopher population; of course I'll set traps and shoot one when the opportunity presents itself. The bigger issue is how I'll approach this whole thing, from what perspective.

I choose the wise-farmer path, at least on most days, and especially on the days when Terra works on the farm with me. There will always be something to hate, and I know my time here is limited. As soon as I step out of the way, the order of things will rearrange and cover my traces, just like an old barn here in the great Northwest, inching toward a return to the earth.

Gopher and Farmer. One.

(I'm still looking into plowing the fields at a depth of six feet, in case this wise-farmer thing is a mistake.) ✍

What I Learned from Gwen

BY CORY CARMEN

After graduating from Stanford with a degree in environmental policy and spending
a few years working on Capitol Hill and in Los Angeles, Cory Carmen returned home
to rural Oregon. Today, she and her husband, David Flynn, are the fourth generation to
raise their children on the family cattle ranch.

I've always been happiest around cattle; it's a trait I
inherited from my father and grandmother. Since I was twenty-four,
raising beef cattle has been my profession. I appreciate the predictable,
indifferent ways of cows, how well they fit in to our mountainous
landscape, and that they require little in the way of human interaction.
But last winter, in the dark corner of a hundred-year-old barn,
my friend Liza, also a beef rancher, introduced me to her dairy
cow, Jewel.

Generally speaking, ranchers don't need milk cows. We like our
work to change with the seasons rather than taking on routine daily
chores. Our cattle are selected for hardiness and thrift, and we count
on them to survive with minimal human intervention. On those
occasions when we do vaccinate or sort them, they can't wait for the
interaction to end.

Dairy animals are strikingly different. Centuries of breeding for
high milk production and easy demeanor have resulted in cows that
are intelligent, sensitive, and friendly.

Before my time, our ranch was home to several Jerseys, Guernseys,
and a few Holsteins. I knew that Grandma had sold milk to neighbors for years. The family referred to the stash from these sales as the
"milk money," and it enabled her to expand her large collection of
antiques and intricate cut glass. When I met Jewel, I thought about all
the stories. For the very first time, I thought, I'd met a bovine that was
interested in more than just getting out of my way.

As much as I longed for my own doe-eyed dairy cow, we already had pigs, chickens, and way too much to do; another animal around the barnyard was out of the question. Yet something strange transpired the following spring. In a herd of beef cows numbering fewer than two hundred, we saw seven sets of twins. Not designed for milk production, Angus cattle often falter at raising more than one calf. Besides, a cow would inevitably lose one of the calves out on the range, forcing it to fend for itself or starve. We coerced a few Herefords into keeping both of their calves, but two orphans remained. So I called Liza, who connected me with my first Jersey, Gwen.

Gwen came from a huge concrete dairy where, at two years of age, she was headed to slaughter for her low production — just three gallons a day. We bought her at a slaughter price of forty-five cents a pound. Her low production seemed like bounty to us and, despite her lack of experience, Gwen raised the orphan calves with admirable dedication. It was like a dress rehearsal for her because as a bonus, Gwen arrived on the ranch two months' pregnant.

As fall approached and Gwen neared her due date, we drove five hours, to Portland, to make the last of four weekend deliveries of beef to our customers. We returned relieved and tired, with checks that combined to make one-quarter of our annual income. But then we found the lifeless body of Gwen's calf by her side. It wasn't clear why the tiny heifer hadn't made it. Gwen hadn't moved from the area of the birth. We took the calf's little body, already partially ravaged by the coyotes, to bury it. I felt an intense wave of sadness. As my affection for Gwen had grown, I dreamed of a Jersey heifer calf and eventually two lovely cows in the barnyard.

The first part of my wish had come true, but I wasn't able to save her. Rage welled up in me. I swore and kicked at the ground, and then, taking a deep breath, recited our family mantra to myself: "It can't be changed. Don't think about it."

Instead, I focused on the task at hand. Gwen faced the very real danger of contracting mastitis, which could prove fatal. The only way to prevent the ailment was to watch her closely and extract all of the milk she produced, at least twice a day.

I felt gratitude to the small animal that required me to spend those quiet hours in the barn and now enabled me to grieve.

Thus began my morning and evening rituals with Gwen. There was a certain serendipity to the sadness. Had her calf died earlier in the year, I couldn't possibly have spared those hours to spend with Gwen. Each and every week prior had been a flurry of phone calls and orders, meetings, and hours of driving. They'd consisted of late nights in my tiny office where I e-mailed beef customers, chefs, and distributors; paid bills; and applied for a larger operating loan. Suddenly, it was November and as the last few checks came in, I realized that for the first time since I began raising cattle, their sum total would be enough to pay off the operating loan. That meant I didn't have to take a part-time winter job to pay the bills. And that meant I could afford the time in the barn with Gwen.

The responsibility of a newly fresh (lactating) Jersey wasn't exactly on a par with what I faced when I'd had my first child, but the sense of duty and pressure was strikingly similar. When I brought home our first baby, I read every baby book I could lay my hands on, trying to prepare for all the problems I was certain I'd encounter. Now I was reading about milk fever and mastitis and ketosis and trying to decide how much and what type of grain to feed. I knew that dairy cows were part of my family history, but my grandma and her vast experience were gone, leaving me in the shed with a bucket and a stool, relics hanging on the side of the barn, wondering what to do. How much milk can I leave? How much grain should I give? How many days of colostrum will there be? And when will my hands stop aching?

Balanced on the stool, I rested my head against Gwen's flank and placed my hands around her long, full teats. The motion felt awkward at first — I pinched with my thumb and forefinger and squeezed with the other fingers, one after the other, until I heard the stream of liquid splashing into the bucket. I repeated the routine morning and evening, each time slightly faster and with more confidence.

The milkings were a welcome reprieve. No phone calls, no e-mails, no lists of demands — just the straw, the warm brown cow with her big black eyes, and the steam that rose from the silver bucket. Some nights I thought about our business and its challenges, but more often I thought of my grandma. Summer had been so busy — with new sales and deliveries and our three young children — that when we heard Grandma's final diagnosis in July, I could only sit by her side, hold her hand, and invoke the longstanding family tradition of setting aside difficult emotions for later.

And that meant not thinking about losing her and what it would mean to us.

I thought of her sitting on that same stool countless times before me. She was the matriarch, descended from wealthy German immigrants. Her father had left the family business to become a rancher and marry the woman who taught in the one-room schoolhouse. My grandma was the first of their four daughters. She grew up like a pioneer child and spoke often of her love of saddle horses. Grandma rode until she was in her late eighties and after that always claimed riding was what she missed most. She was a loyal and devoted grandmother who taught me how to make big roasts and mashed potatoes and gravy. She canned our fruits and vegetables and delighted everyone with her lemon pies. But these were skills of necessity. Her passion was for animals and, even in her nineties, she would drive the feed pickup for us in the winter because seeing the cows gave her such joy.

When I came into the world, Grandma was already sixty. She lived in a different time, and after my own father passed away, my grandmother was my connection to our family legacy. She and I spent evenings looking at old photos of her childhood. She told me about the milk cows, the pigs, and the chickens that her mother raised. In those days, they kept the steers on grass until it was time to put them on the train to Portland. She had decades of knowledge and experience that tempered the hard times we encountered. When we lost two of our best cows, heavy with calf, because they fell upside down into a ditch, she told us about a morning decades earlier when Grandpa had awakened to find six dead cows in the same ditch.

Even though she no longer worked on the ranch every day, we felt her presence everywhere. Our cows came from her cows. The barns and sheds we used were once hers. And the house where we live is where she raised my father and uncle. Her lifetime of dedication and perseverance had made it possible for us to ranch, and she passed on to us, along with the assets, her expectations: hard work, taking hardship in stride, pride, faith, charity, and loyalty.

I knew I was in charge of the business of running the family ranch, but her death pushed me to the emotional helm of the ranch, a role that felt overwhelming. I made decisions and implemented changes with confidence, but my assured manner was a thin veil that barely masked the intense pressure I felt to keep the ranch going, and the reoccurring uncertainty that I could pull it off.

One evening before Grandma died, she said, "You do things a lot like my mother." We had stopped feeding our cattle grain and instead were selling shares of beef to city dwellers who appreciated our connection to the land. We'd also picked up some wholesale accounts and were supplying beef to several colleges and universities, as well as to a large hospital. Grandma's observation reassured me that, through all the changes, Dave and I were stumbling down a path that had been walked before us.

At Gwen's side, the streams of cream turned to drops and for the first time I allowed my tears to fall to the straw. I felt gratitude to the small animal that required me to spend those quiet hours in the barn and now enabled me to grieve. As I carried the bucket of milk back to the house, I wished Grandma could have been with us a few months longer.

A day earlier, I'd taken the ripened cream from the kitchen counter and poured it into the Kitchen Aid mixer she gave us for our wedding. The cream rose and whipped and flattened. And stayed flat, never producing butter. I looked for advice online, longing for the customized guidance Grandma would have given. She would have known how to make butter from a Jersey eating our grass and our grain at our exact elevation. There was no substitute for what she knew and what I lost when I lost her. Even though I wanted her to see me pay off our loan, even though I longed for that lesson in butter-making, even though I have many new questions to ask her, I cling to the answers she already gave me, especially this one: We are on a familiar path and, through stubborn perseverance, we too will make our living on the ranch. ✍

The Ambush

BY CARDEN WILLIS

Carden Willis runs A Place on Earth CSA Farm in Turners Station, Kentucky. He and his wife, Courtney, became the proud parents of Clark on Thanksgiving Day, 2010. They are blessed with a wonderfully supportive community and a sublime life — ecstasy, agony, and all.

Yesterday was the first rainy day in many moons. No fieldwork could be done, no chainsawing. But I knew I couldn't leave the farm for long. Not with Courtney more than forty weeks' pregnant, teaching at school, potentially calling at any moment to say, "Come and get me!"

I was gone only a couple of hours, and as soon as I got home, I made a beeline to the answering machine. No dramatic blinking lights, no good tidings. I recalled that Tierra, our dog, was acting a little odd as she welcomed me home. I called her to come inside. She didn't.

Strange.

Stranger yet was the sight out the window from our bedroom: The chickens, spread across their yard as usual, were not moving. No, those couldn't be chickens. They were . . . lumps of . . . something. Something else. Even from fifty yards, though, the truth was starting to sink in, like blood into straw.

I followed Tierra to the scene, as if in a dream, and several possibilities suggested themselves to me. Each was successively dismissed, crushed by reality. The chicken-size shapes on the ground were not giant leaves. They were not chickens taking deep dust baths in the rain.

As I got closer I started to feel dazed; I wanted to wake up, to open my eyes, to turn over in bed, to try a different dream. The silence and stillness were stark and surreal, the quietest space I had ever been in. Not a cluck, not a twitching feather. No alarm clock clicked on to save the day. The dog and I, together at a standstill and stunned, were in a sort of netherworld, stuck between suffering and acting.

My feet somehow carried me through the carnage. The circumstances of the massacre piled up: the fence pulled down, chunks of chickens missing, scalps of feathers strewn about.

I rolled the stones away from the door of the coop. The floor of straw litter was gray and black and wrong. More stiff, broken bodies. The bustling metropolis of a few hours before had turned into a ghost town. I had let down my flock. I believe Tierra felt this too.

One of our two roosters, Bono, had been our constant companion for almost five years — years of crowing his heart out every day like a rock star. I could almost feel his last throes of rage and honor. I could feel also the mass panic and terror that swept through the crowd of sixty hens. But mostly I felt the unsettling calm after the storm and my broken promise to protect my family.

How do I tell Courtney? What if she sees this? What do I do?

As I left the coop, I heard a movement. A shocked Australorp nervously tried to balance atop a feed can in the corner. I knew what she had seen, and my heart broke for her. I wanted to explain: a big mistake. I'm so sorry. Whose eyes could ever recover from watching her family murdered, one by one? What would "I'm sorry" be worth?

My mind moved to the impending birth of the fully formed life in Courtney's belly, the polarity of life and death, the single heartbeat that divides the two, the eggshell fragility.

The chickens, spread across their
yard as usual, were not moving.
Even from fifty yards away the
truth was starting to sink in, like
blood into straw.

On April 16, 1998, during spring break of my junior year in high school, I was visiting my grandparents in Nashville. My grandfather was out grocery shopping while my grandmother and I watched an old movie. By the time we figured out that something was happening outside and had made it to the bathroom, the event was over. The tornado, with winds up to two hundred miles an hour, had come and gone. I saw a new world as I emerged from the house. All was eerily still and uncannily changed. The world, at that moment, was otherworldly.

Trees were bowled over and strung about like children's toys, including one on top of my dad's Camry, which I had driven, and many others crisscrossing the driveway and Rosebank Avenue. With all the tall old trees down, the quality and the quantity of light in the sky were different. The carport was gone, later to be found in a backyard several houses away. It's in a rare blink of the eye when reality is profoundly altered, but when that happens, there is no amount of rubbing your eyes that will recover what has vanished.

I don't trust my memory. I'm curious how long the stillness lasted. However long, it was enough to buoy my faith in black holes. At some point, though — after an infinity — consciousness returned. From the great void resumed the familiar sound of chain saws, men and women talking, crying, strategizing. The pieces were picked up but could not be entirely put back in place. It was some time before the roads were cleared, and eventually we heard from my grandfather. He was all right. That much reality was restored.

Luckily, now, our neighbor Bonnie was home yesterday afternoon. The task before me — alone — seemed huge. When she opened her front door, I only had to say "I need your help" and she sprang to action. Together we gathered up all the birds, dead, dying, and alive but shocked. In the end, we recovered forty-four bodies and thirteen hens still alive.

Bonnie's husband, John, helped me bag up my old friends. Then the three of us went on a search party through the wet weeds and blowing rain. I found a couple of feathers leading to another neighbor's place, and John and Bonnie saw the Mayses' guilty dog with a chicken in its mouth. Tierra, with her keen nose, discovered a body in

The Ambush **139**

the tall grass behind their house. By dark, the case was closed and we were drenched.

Later in the evening, Bonnie came back and we tended to the living. She packed antibiotic cream into the maimed flesh and puncture holes while I held the poor creatures in place. In times of deepest distress, we survive by the power of courageous human kin who don the rubber gloves and fighting spirit and care for the living. What would we do without angels?

I believe in the opening of eyes. I believe in instantaneous transformation. I believe the very walls around us can collapse and, as long as we have community, the beautiful human spirit will rise up from the rubble, heal the broken bones, and rebuild, one precious life at a time.

Yesterday, when all but one of the surviving chickens had turned up, I was heading back to the house from the coop when I saw a shape in the distance. Soaked and with a few feathers missing, there stood Little Richard, our other rooster, bewildered, making his way to his ravaged home. My heart leapt when I beheld him. A humble triumph. Thanks be for survivors. ✍

Notes from a Novice Horse Farmer

BY ALYSSA JUMARS

With her Belgian draft mare, Ray, Alyssa Jumars farms on the east side of
Washington's North Cascades. She's still on the steep part of the learning-to-use-
draft-power curve and imagines she'll be there for quite a while. At the moment,
there's no place she'd rather be.

If I had known then what I know now, I wouldn't be trying
to farm with draft power. But I'll readily admit that buying my first
team of draft horses was the very best mistake I've ever made.

It sounded like a good idea at the time: What could be more
sustainable or more romantic than farming with horses? I'm of
the mind that there are many old-time manual skills on the chopping block that would be well worth preserving, and knowing how
to farm with a power source other than petroleum is one of them.
Two years ago, then, I let myself be persuaded by a former business
partner into buying a team of beautiful, six-year-old, recently broke
Belgian mares.

The first month, we were giddy. Every chance we got, we would
hook the mares to our newly acquired wagon and take off through
the hayfields and into the neighboring orchard, sun in our faces, lines
in our hands, the smell of dust and horse sweat all around us. There
was the creaking of running gears, the jingling of trace chains, the
rhythmic *clop, clop, clop* as we bumped along rutted farm roads. We
would head for the far end of the orchard, where the road ran straight
for a quarter mile, kiss to the girls, and push them into a trot. Bracing
ourselves against the front of the wagon, we would glide past the rows
of pears and apples.

My favorite experience, though, was our first time plowing. We were turning under a section of winter peas and rye, getting it ready to plant a big patch of black-seeded sunflowers for chicken feed. My partner and I took turns steering the walking plow and driving the horses. For the first half hour, we wore ourselves out wrestling with the plow and trying to keep it straight and in the ground. Only when we were almost spent did we begin to realize that the plow drives itself and that you merely have to guide it subtly one way or the other to keep the landside suctioned into the unturned sod and the bottom of the share suctioned into the dirt. It was so simple, so elegant. I'll never forget being behind the plow, feeling the raw power as it surged from the girls and watching the sod bust open and fold into ribbons before me. I had never felt so alive as behind the great ass of a draft horse.

In the beginning it all seemed so easy: The horses were gentle, obliging, and trusting. We plowed close to three acres with them and drove countless miles on the farm roads. But the honeymoon would soon come to an end. In our novice hands, the training they had received from the breeder began to wear off, and the horses started to test our authority in small ways. They would turn away and walk off when we came to halter. They'd try to walk ahead when we led them from the pasture to the barn. They'd take a step to the side when we slid the harness over their backs, or raise their heads when we bridled them.

Slowly but quite surely, their gestures became more exaggerated. At the sight and the jingling of harness, they would swing ninety degrees, twitch as we put the hames over their shoulders, and generally act terrified of the jumble of leather they had seen a hundred times before. When we bridled, they would pull back on the lead rope as if their lives depended on it. Our reaction to this nonsense was to assume that the horses were fearful; really, they were just looking for creative ways to avoid us and to find out how much they could get away with. If one of the horses began pulling back when we brought the bridle to her nose, we would take away the bridle, talk soothingly, stroke her neck. What we unwittingly taught them was this: Throw a fit or act terrified of something and we will make it go away.

This is the single most dangerous lesson a horse can learn, and we taught it well. Our horses learned that we were pushovers and that they could get away with holding their own opinions and, ultimately, making their own decisions. As a result, we had three major wrecks.

There is almost nothing more terrible than that second when you realize you're no longer in control, that you're not going to be able to hold them back with the lines, that these animals you so dearly love are about to run themselves blind through barbed-wire fences, cut themselves on the implement they're dragging, get tangled in harness, or worse. And it's entirely your fault: for not reading the signals they were giving you; for not insisting that you were the boss; for asking of them more than they were psychologically ready to do; for not paying close enough attention and missing that split second when you had an opportunity to correct the disaster now unfolding in front of your eyes.

Our first wreck happened with a riding cultivator. We had picked it up at the Small Farmers' Journal auction that April, and we couldn't wait to try it out. We were doing a fifty-five-share CSA, and we had about four acres of row crops that we were eager to weed any way other than by hand. A riding cultivator straddles a bed of vegetables and drags adjustable sweeps that are spaced at exactly the right distances to weed around your one, two, three, or four rows of crop. You sit on the machine and the horses walk in the paths on either side of a bed.

I had never felt so alive as behind the great ass of a draft horse.

After spending a dozen hours in the shop, scraping out and replacing all the fifty-year-old rock-hard grease, we rolled the cultivator out to the barnyard. We tried to think of ways to gradually introduce the new tool to the girls. One of us walked them around and around the cultivator so they could get a good look. The other lifted up the tongue and jiggled it so they could see that it moved and could listen to it rattle and clank. Then we hooked them up and made them stand for a few minutes. We took just a few steps down the farm road and stopped. A couple of more steps and stopped. They acted like they were quite comfortable with the situation, so we proceeded to the field and made a dozen passes on a bare bed. Everything went great, and we called it a day.

The intensity of working with the horses exposed the fragility of our relationship and our incurable mistrust of each others' judgment.

The following morning, we brought in the girls, harnessed them up, and began hitching, confident that they were now accustomed to the new machine. We went about things with a good deal more speed and a good deal less caution than we had done the day before. I held the lines while my partner hooked the yoke and the traces. He had one of the girls hooked in and was walking around to hook in the second horse when the second horse swung her back end around a hundred and eighty degrees.

We learned a cardinal rule of working with young teams and new tools that day: Don't ever try to hitch without using a butt rope. A butt rope essentially ties their back ends together, preventing them from being able to fan and turn inside out. Our second horse was now turned inside out, the lines were reversed relative to the other horse, and she was standing looking directly back at this new machine — that, as it turns out, she was in no way fully accustomed to or comfortable with.

She panicked, the other horse panicked, and away they went with the cultivator. They lost the cultivator pretty quickly, but it wasn't until they ran through an old fence, going on either side of a post, that they broke free of each other. One horse got tangled in the fence and the other headed for the main road.

Our second runaway took place a few weeks later. After that first wreck, we decided we would stick with tools we knew they had used on the farm they came from until we had all rebuilt our confidence in each other. We decided we would hitch them to a spring-tooth harrow and bust some clods in a field that had recently been disked. What neither of us understood was that an old tool in a new place with a new driver is, to a horse, effectively a new tool. We assumed that because the horses had harrowed before, there would be no problem.

We didn't think to take any precautions, like trying out the harrow in a small, fenced pasture. The horses took three steps, felt this strange new weight behind them, panicked, and took off again. It hadn't even crossed my mind that they would run and that I might want to be ready to use a lot of extra line pressure to stop any momentum before it built up. By the time I reacted to the situation, a split second later, it was too late.

Amazingly, the young mares never suffered any major physical injury in either of the runaways. But their confidence in harness was irreparably damaged, and they had learned how to run. By all accounts, my attempt to farm with horses had been an utter failure. I had ruined a good young team, and if I had been any less lucky, I could have killed them.

I tried to convince my partner that we should bring the girls back to the breeder, that we had no business owning horses, but he was sure he could find a way to make it all work. Maybe he has, I don't know. I left the farm at the end of that summer. It was the horses that split us apart; the intensity of working with them exposed the fragility of our relationship and our incurable mistrust of each others' judgment. When I left, I felt like I was abandoning the girls, and for months afterward I tried to talk my old partner into giving them up. But in the end, I had to let go and move on, so I sold him my share in the team.

At that point, I probably should have returned to tractor farming. But I was hooked. There was something about partnering with another creature that I could now never give up. The sheer implausibility of asking an eighteen-hundred-pound beast to do exactly as you ask, in a language consisting of four words and a hundred subtle gestures, blew my mind. So I went in search of mentorship. I found some, though not as much as I would have liked.

Finding someone with the ability, the patience, the inclination, and the time to teach you everything you want to know about drafts is a challenge in and of itself. I spent a lot of time observing other farmers' workhorses, but not a lot of time with the lines in my hands. I slowly came to realize that I could learn bits and pieces from differ-ent people, but that the only way I was ever going to get a chance to practice and put it all together was to have horses again.

Learning to farm with horses is a huge catch-22: You shouldn't own a horse unless you're experienced, but you're never going to get that experience until you have your own horse.

My second go at buying horses, I was more deliberate. This time I decided I wanted just one horse. In my mind, one horse would be more manageable than two. (As it turns out, it's not. It's just different. I happen to like it better.) And I wanted an old horse, well past the piss-and-vinegar years. I didn't care what breed, what gender, what conformation; I just wanted a horse with a lot more experience than I had. I got lucky. Some farmer friends decided to sell me an older, very experienced mare. And they generously decided to give me a num-ber of one-on-one lessons with the horse before I hauled her the four hundred miles home.

It's been seven months since I got Ray, and I'm still taking it slow. I've finally gotten it through my head that I don't have to learn to plow, disk, rake, mow, cultivate, harrow, and log all in one season. It takes time to build a working relationship, and it's always more important to make sure that every experience Ray has is positive and safe than it is to accomplish the task I set out to do.

I've been around a handful of other beginning teamsters, and I think most of us suffer, initially, from mild to severe cases of ignorant bliss. We think horses are lovely, docile animals; we hatch elabo-rate plans for all the things we'll do with them; and we simply don't understand how wrong things can go or just how much attention has to be paid at all times. On top of that, we tend to be an especially ambitious, passionate, and pig-headed lot. I'll be the first to admit that

I invariably bite off more than I can chew; I don't always take advice well; and I usually think I can figure things out on my own.

Learning to work with horses is, first and foremost, a lesson in humility. I've had to learn to ask for help and to admit when I don't know what I'm doing. I've had to learn that it's not up to me to determine what gets accomplished in a working day or in a farming season, but that it depends on the relationship I've created with my horse and the time I've taken to build her trust and to demand her respect and compliance.

And I've had to learn to focus. When I'm working with Ray, I can't be thinking about the salad bed that needs weeding, the drip tape that needs patching, the chef who needs an invoice. I must be completely present. I have to be hyperaware of what's going on around me, and yet focused on reading Ray's body language. Are there any animals — dogs, horses, people, other livestock — that are going to distract or become a nuisance? Are there any dips in the topography that are going to make the implement lurch forward? Is there anything we might snag on? Is the angle of draft correct or do I need to adjust the length of the traces? Is anything going to come loose on the tool? Is the harness rubbing anywhere? Is Ray paying attention to me or do her ears indicate that her attention is elsewhere? Is her neck raised, showing general concern, or is her neck lowered, indicating that she's comfortable and at ease? Am I clear of the implement if it hits a rock and swings sideways? Is Ray panting; should I let her catch her breath? Is she trying to rush through a particular section of field or through a turn? Why?

And at the same time, I have to breathe deeply, relax my shoulders, speak steadily, and project calm and confidence. I've never had to do anything more challenging. And I've never loved any challenge more. ❧

Moral Clarity through Chicken-Killing

BY SAMUEL ANDERSON

Samuel Anderson is the livestock coordinator at the New Entry Sustainable Farming
Project, a Massachusetts-based farmer-training nonprofit. He grew up on a sheep
farm in Ohio and has worked as an agricultural journalist and a magazine editor.
Sam graduated from Kenyon College and received his master's degree in urban and
environmental policy and planning from Tufts University.

**Standing under a stainless-steel tangle of shackles
and blades,** I began to see my chicken-killing experiences in a
different light. I was at the International Poultry Expo in Atlanta, an
immense gathering of representatives from every corner of the poultry
industry, perusing the exhibits set up by various poultry-processing-
equipment manufacturers, trying to cover the word sustainable on my
name tag so the company reps wouldn't brush me off.

More specifically, my nametag read NEW ENTRY SUSTAINABLE
FARMING PROJECT, the Massachusetts nonprofit where, in the course
of less than two years as livestock coordinator, I had gone from bright-
eyed farm kid and local-foods enthusiast to practiced chicken killer
and an authority of sorts on mobile poultry-processing units. At New
Entry, we have been trying to fix a ubiquitous problem in the US food
system: the glaring lack of local, small-scale slaughterhouses and
meat-processing facilities. Our solution: a mobile poultry-processing
unit that we trained farmers to use, so they could kill and process their
own birds for sale to local markets.

I trekked down to the poultry expo because, as part of a regional
poultry science training project, my name had been drawn out of a hat to
get an all-expenses-paid trip to the world's largest poultry-industry event.

My first thought upon arrival: We're not in Massachusetts anymore.

Working with small-scale poultry farmers in the Northeast and having raised my own poultry growing up, I was used to small, pasture-based, independent chicken production — the kind you can refer to with a straight face as "farming." At the International Poultry Expo, "farming" has a different look. In the big poultry industry, a chicken farmer is someone raising thousands of broilers on contract in a huge poultry house. When the birds are mature, a crew trucks them off to the giant, centralized processing plant — owned by the same company that holds the contract, such as Tyson or Perdue — and by the time the farmer has fired up the tractor to start cleaning out his empty poultry house, his chickens are already being shuttled down the highly mechanized (dis)assembly line that will speedily convert his live birds into uniform carcasses to be shrink-wrapped and sold all over the country.

For the farmer, the killing and processing of his birds is out of sight and out of mind.

It's a foregone conclusion in the big poultry industry that, as the University of Georgia's broiler information web page bluntly puts it, "Having an independent broiler-growing operation is no longer feasible." In fact, according to the site, there is actually no such thing as an independent chicken farmer anymore; "approximately 99 percent of all broilers are produced under contract, with the remaining production occurring on integrator-owned farms [those that are owned by the same vertically integrated company that owns every stage of production, including the processing facility and the retail brand]." The argument is that small-scale production can't compete with the low prices of the ultra-efficient industrial operations.

As I wandered among all the menacing machinery at the expo — chicken-gassing chambers as tall as houses; conveyor belts that drag chickens shackled upside down through an electrical stunning bath and a razor-edged gauntlet; an automated contraption that zips chicken carcasses around like a carousel to pluck out their viscera — it dawned on me that it was these very machines that had made our mobile processing unit in Massachusetts so necessary. The more consolidated and mechanized meat production becomes, the fewer small-scale, local processing facilities exist and the fewer independent farmers there are. The expensive, high-capacity equipment not only removed every ounce of intimacy from killing and processing chickens, but it also drove independent processors out of business, forcing independent poultry producers to get big or get out.

The processing bottleneck is a perennial problem for small meat producers all over the country. Here in the Northeast, processing woes are poised to overtake regulations and the weather as the favorite farmers' griping topic. They have plenty of reason to grumble: Massachusetts has only two small USDA-certified slaughterhouses. Poultry producers — and those who would like to become poultry producers — face an even narrower bottleneck. Before the construction of three mobile poultry-processing units (MPPUs colloquially, or "chicken-slaughter-houses-on-wheels" if you're not into the whole brevity thing), the only legal poultry-processing option available was to build, operate, and obtain state licensure for a stationary on-farm slaughter facility. Most farmers find that too expensive.

I've talked to plenty of farmers who have decided against raising poultry, dismayed by the barriers to getting their birds processed. Others attempt to operate under the radar, quietly processing their own birds or taking them to a non-inspected processor, then selling directly to a trusted customer list. Even if these growers escape the notice of regulators — and they usually do — you won't see their birds at a farmers' market or restaurant, and certainly not at a grocery store. For lots of small producers in Massachusetts, the mobile processing units are the only legal and affordable processing option available. But there are strings attached, not least among them the fact that the farmer has to be hands-on when it comes time to kill, pluck, and eviscerate her birds.

As it turns out, it might be worth it. There are still independent chicken farmers in the world, and they've proved that it can pencil out. In 2010, three Massachusetts producers utilized a mobile poultry-processing unit to legally process their chickens. Each raised, hand-processed, and sold between eight hundred and twelve hundred, all grown on pasture. Through farmers' markets, restaurants, and presales directly to consumers, these birds fetched from four dollars and fifty cents to six dollars per pound. Some of those chickens topped thirty dollars each.

> You're killing a living creature and you're not quite sure what gives you the authority to be doing this.

Compare that to contract chicken farmers. They get paid between 3.8 and 4.6 cents per pound of live weight. That means that a particularly efficient producer might gross a whopping twenty-five cents for each bird. During the ten to fifteen years it takes a contract producer to pay off the hundred thousand dollars in up-front cost of building and outfitting a poultry house that meets Tyson's or Perdue's standards, the farmer needs to grow well over a hundred thousand birds a year just to net five thousand dollars.

The savvier independent chicken farmers are making a hundred and twenty times more than the gross per-bird return of a contract broiler. One of these producers calculated her annual net return at somewhere around ten thousand dollars — raising fewer than 1 percent of the birds it would take a contract grower to get there.

The independent farmers had to kill each and every one of those birds themselves. This is no small matter — which is why when New Entry was hiring a new staff member to help with the MPPU project, Jennifer Hashley (my esteemed boss and a hugely accomplished Greenhorn), drew on her own experience as a chicken producer and MPPU user. The first qualification: "BA or BS preferred." The second: "Must be willing to teach farmers how to process chickens." Note the key word there: "Must be *willing*," not "Must be *able*."

I convinced Jennifer that I was willing, and then I convinced myself. After all, having grown up on a sheep farm, it wouldn't be the first time I had willfully and purposefully killed an animal. I had killed a deer with a shotgun, a muskrat with a steel-jaw trap, and a groundhog with a cinder block; and though we didn't slaughter them ourselves, by the first grade I was well aware that when I gave a newborn lamb an orange ear tag, I was sentencing it to market and an early death. The gross-out factor wouldn't be an issue, either: I had plucked chickens; hooked and filleted fish; shoveled manure; and castrated, vaccinated, and cut the tails off sheep. I had, on more than one occasion, strapped on a pair of surgical gloves, reached into a ewe's backside, and pulled out a slimy orange lamb.

All of that helps, but when you find yourself holding a knife to a chicken's throat, you may discover that you haven't quite covered all of your bases. Yes, you will need to have learned the actual technique — how to place the bird in the cone, how to hold the knife, how to apply the stun, how to make the cut — but in truth, the physical act of killing chickens is easy. The emotional act is more challenging. You may be able, but are you willing?

The first chicken you kill probably won't be the most difficult. You'll direct your nervous energy toward focusing on the technical details, making sure you're going through the proper motions, and a moment later you'll realize that you did it and that it wasn't so bad after all. A wave of relief and adrenaline will carry you from there, and you might feel pretty good about yourself for pulling it off. The most difficult bird will come later, when you no longer need to keep your mind trained on the motions and it begins to wander, and you finally process what's going on here: You're killing a living creature, a whole crowd of them, and you're not quite sure what gives you the authority to be doing this.

The easy thing is to brush off those thoughts, put them out of mind and keep them there; but taking on the moral and emotional questions is, I think, essential. This is what separates the independent, earthbound chicken farmer from the mainstream, impersonal-by-design broiler industry. The big broiler-supply chain goes to great lengths to get everyone off the hook — the companies, the producers, and most of all the consumers. When you process your own birds by hand, you aren't letting yourself off the hook. As a customer, by volunteering to help out your farmer on processing day, by going to the farm and learning how it's done, or just by the act of considering how the chicken you're buying was processed, you aren't letting yourself off the hook. And the next time you think about buying a nameless chicken at the grocery store, you'll ask yourself, as I did: I'm able, but am I willing? ✄

The Gift

BY KATIE GODFREY

Katie's first farming experience was on a biodynamic farm at the Michael Fields
Agricultural Institute in Wisconsin. She wrote "The Gift" during a writing residency
at the Wisconsin-based Wormfarm Institute, a nonprofit that works to integrate
culture and agriculture. Katie is currently farming in the Driftless Region of
Wisconsin, an area historically missed by glaciers, preserving the unique topography
of looming bluffs and winding rivers.

I was wearing a tent.

It was my first day of beekeeping and the bee suit that I had borrowed was almost as wide as it was tall. It was ninety-five degrees
outside and humid, but I was determined not to get stung. It's not that
I have a fear of bees — it's more a fear of being chased and stung by
a lot of them at once. After reading that each hive houses an average
of twenty thousand to sixty thousand bees, I wasn't about to take any
risks. In the safety of my mentor's truck, I clumsily stuck a veil on
my head and double-checked the ties around my ankles to make sure
nothing could crawl up my pant legs. The borrowed elbow-length
gloves had grown stiff with years of wax and honey drippings, making
it impossible for me to bend my fingers. I felt like a messed-up version
of a school mascot headed to get the crowd going for certain defeat.

I was enrolled in a farmer-training program in southeastern
Wisconsin, where I worked with four other interns on an organic-
vegetable farm. Most of us were recent college graduates — urbanites
who wanted to get a taste of the local-food movement at the source.
As part of the program, we worked with a new mentor farmer in the
area each month. In May, I worked with a woman who keeps dairy
goats. Now that it was June, I was working with Dan, the beekeeper.

I fumbled my way out of the truck and noticed that Dan was
already busy working with the little critters. I also noticed he wasn't

wearing a bee suit. Or gloves. He waved me over to one of the hives and I nervously eyed his veil, which was tied loosely enough to let in a few rogue bees.

As I approached, the air grew thick with the song of a million paper-thin wings beating in unison.

"Don't worry about getting stung," Dan said without looking at me. "They'll bump you first as a warning. They don't want to sting you, because they'll die if they do." I felt slightly better knowing this.

Then Dan continued, "However, if you do get stung, an alarm pheromone is released, which attracts more bees and they'll chase you until you're good and gone." I shivered and took a step back.

I was grateful that Dan refrained from commenting on my absurd uniform, but he did inform me that I had put my veil on backward. He quickly spun it around and tied it to my suit.

He has a kind face, lined with wrinkles that deepen when he smiles. Dan grew up here in the town of East Troy and married his high school sweetheart before taking over his dad's plumbing business. In the car ride over, he talked proudly about his work, but his eyes really came alive when he started talking about his true love: the bees.

"We have a lot of work to do today," he said, surveying his miniature village.

I glanced at the gangs of bees moving in and out of their homes and meekly told him I was up to the challenge. Almost every hive was composed of two white boxes stacked on top of each other, like little white apartment buildings.

"Those are the brooders," Dan explained. "That's where they lay the eggs."

We were looking at about twenty hives parked in the shade of trees at the end of a long field. Dan has hives scattered all over town, mainly on farms. I glanced beyond the field and noticed an old woman pulling weeds in her garden — it was his aunt Ruth. She had moved to the area from Germany in the 1940s and brought with her biodynamic-farming principles, a type of organic farming that treats each farm as a living system. Dan, who clearly grew up entrenched in these principles, stressed that bees are an integral part of this system.

Our task for the afternoon was to look for eggs in each hive, to make sure the queens were fulfilling their purpose. If we couldn't find any eggs, we would have to find the queen herself to make sure she was still alive; otherwise, the colony would most likely die.

Dan lifted off the top of the first hive and we looked down on ten frames, like ten long books in a fallen shelf. The brooder was swollen with bees moving vertically throughout the thin spaces between each frame. My eyes widened as he plunged his naked hands into the open hive and gingerly lifted out a frame. At least twenty bees crawled over his sun-baked hands as he nonchalantly began demonstrating how to find the eggs. Horrified at the scene, I interrupted him to ask why they weren't stinging him.

He looked surprised, as though it's perfectly normal to use bare hands to handle insects equipped with weapons on their butts.

"Oh, by now we've built a relationship with each other, haven't we?" he said rather sweetly to the bees that were examining his hands in search of nectar. "They're used to me invading their space once a week. I know better than to come on a cold rainy day, when they're irritable. On a sunny day like this, they ignore me."

Even though I was wearing what was practically body armor, I silently hoped they would ignore me too.

I watched Dan as he held up the frame to the sunlight to search for eggs. The wax film that was set into the wooden frame was partially filled out with hundreds of hexagonal shapes, but they were empty: no eggs to be found. He replaced it and retrieved the next one, then repeated this process two more times until, finally, there they were: the minuscule white lines that were bee eggs, each one no larger than a comma. The next frame housed larvae and a few odd-looking cells that stuck out from the frame like warts.

"What are those?" I asked, repulsed by the bumps, which resembled boils marring an otherwise beautiful face.

"Those are drone cells," Dan said, "eggs that turn into males. The hive is almost entirely run by females, you know." He glanced at me. "Out of all of these bees, only a couple of hundred are male. All the worker bees are female. They do everything but lay eggs and mate. The drones," he said, smiling, "are kept around simply for mating with queens from other hives. When they're not mating, they sit around and gorge themselves on the honey and pollen that the females worked so hard to store. But they die soon after they mate. And if they fail to mate, they're thrown out of the hive in the fall and they die anyway. The women don't want to keep them around for the winter." He chuckled at the thought.

Dan placed the frame back in the hive and changed the topic.

"Well, we know this hive is active," he said, "so we don't need to find the queen." Satisfied, he replaced the lid on the hive and we moved on to the next one.

"What happens if we don't find any eggs but the queen is still there?" I asked.

Dan got a gloomy look on his face. "Then I have to perform the depressing business of killing her, and introduce a new queen, although often the colony will start creating a new queen on their own."

I was confused. He continued: "They choose some larvae to feed what's called the royal jelly. This jelly creates a new queen that fights the old queen to death in order to claim her throne. These queen cells look different from normal eggs cells, too — they stick out on the bottom of the frame like raindrops, like they're trying to hide the fact that they're planning a mutiny."

My eyes widened as he plunged his naked hands into the open hive and gingerly lifted out a frame.

Dan let me look through the frames on the second hive while he recorded our previous findings in a little notebook. After about ten minutes of painstakingly removing each frame and trying not to crush any bees with my stiff gloves, I told him, embarrassed by my inability to see the little white commas, that I couldn't find any eggs. Dan did a quick search and I was surprised to discover that I was right — there were no eggs in this hive. We would have to look for the queen.

My stomach jumped. I wasn't entirely sure what I was looking for, except that the monarch would be larger and longer than the other bees. The worker bees and drones seemed to know what we were up to; they started frantically moving around the hive as we dug deeper, pulling out all the frames, our eyes inspecting each body.

Then, suddenly, there she was. Her elegant black figure was moving discreetly around the bottom of the hive. I pointed her out to Dan.

"Good eye!" he said approvingly, though I was sure he saw her long before I did. I took the compliment anyway, even as I felt a twinge of regret about the fate that awaited her. I didn't want to have to watch Dan kill a queen, and I was relieved when he pointed out the lengthy cells dripping from the bottom of the frame. The colony would take care of this problem on its own.

We repeated the process of pulling out frames and searching for eggs with each of the hives. After a while I realized that something was missing. A really big something.

"It's too early in the season to have a lot of honey at this point," said Dan, anticipating my question. He pointed to a hive behind him. "See that third, shorter box on top of the two brooders? It's called a honey super. You place a barrier between the brooders and the supers that's just large enough for the worker bees to slip through but too small for the queen, so she can't lay any eggs up there. That's where they deposit the honey and encase it in wax. The colonies that really thrive end up with four or even five honey supers by late August."

His dreamy voice trailed off. "We'll save that one for last," he said, and we reluctantly pulled ourselves away from the hive.

We worked side by side for another hour, searching for eggs and queens, eggs and queens, until at last Dan called me over to the hive with the honey super on top. I felt a rush of greedy excitement when a sudden thought occurred to me.

"Don't they need honey to survive the winter?" I asked. "I thought that's why they made it in the first place." I was concerned. Were we stealing from the bees?

"Oh, they make more than enough honey to survive," Dan assured me. "The rest of what they make is like a gift to us. We give them a place to live comfortably, and in return they leave us a sweet treat."

With that, he scraped out a piece of honeycomb that had stuck to the underside of the lid and handed it to me.

"And they won't get angry?" I asked. Without waiting for a response, I cautiously slipped my hand out of my glove and grabbed the dripping piece of comb, fat with honey. I turned it around in my hand to make sure it was clear of bees, lifted my veil, and then bit down. The warm honey drizzled into my mouth and the wax melted between my teeth. I grinned at Dan and briefly imagined the day when I would harvest honey from my own hives.

Bees hummed around my exposed head, but all of my fear dissolved as I stood there in bliss under the scalding sun. This was a gift worth a thousand stings. ✍

How Animals Sell Vegetables (and Make You Tired)

BY LYNDA HOPKINS

With her husband, Emmett, Lynda Hopkins owns Foggy River Farm in Healdsburg,
California. They sell sustainably grown produce and eggs through their CSA
and at farmers' markets. Lynda is the author of *The Wisdom of the Radish,* a book
chronicling her farming adventures.

Sedona's kids were sitting low in her belly.
Overnight, the ligaments around her tail head had softened, then
disappeared entirely. I was checking her every hour.

At eleven in the morning, when I came into the barn and sat down
beside her on the straw, she got up and laid her belly across my lap.
She didn't make much noise — just a few ladylike grunts, verging on
coughs — as she started to push.

I called Emmett, my husband and farm partner, on the phone; he
was across the road at the vegetable field. I told him Sedona was starting to get serious.

"Okay," he said, "I'm just finishing up some seeding. I'll come back
in a few minutes."

Right after I hung up the phone, an amber bubble appeared. I
called Emmett back.

"Come now!" I said. "I can see the hoofs."

Emmett arrived just as the first kid, a gold-and-white miniature
replica of Sedona, slid into my hands. I suctioned the mucus out of the
kid's nose and mouth, checked under the tail —a doeling! — and put
her in front of her mother.

Ten minutes later the second kid was born. In the adjacent stall,
Tuxedo — our inimitable herd queen and Sedona's best friend — was

wondering what all the fuss was about. She put her front hooves up on the hog panels that separated her from her friend, craning her neck to get a look at whatever on earth was happening. When the wobbly wet kids tottered over toward her, she snorted. You could see the shock written on her face: Where did they come from?

And although Tuxedo wasn't due for a few days yet, she got right down to business. Chalk it up to jealousy, sisterhood, hormones, or all of the above: By that evening, Tuxedo's kids had dropped into position and her ligaments were so soft that I could reach my fingers all the way around her spine. More noticeably, every time I left the barn, she screamed bloody murder.

So I spent the night with her, and the following morning Tux successfully ushered her own pair of baby goats into the world. Two sets of doe twins: Our girls had done well.

After she'd passed the placenta and the kids had each gotten their first drink of colostrum, I glanced across the stall at Emmett. Stuck to his face was a huge, stupid, dumbstruck grin. I realized I had one, too, and it occurred to me that we enjoy spending time with our animals for many of the same reasons our customers do.

The quintessential farm — that romantic ideal that rests in every child's heart — isn't just a silent field of vegetables. It's the cock crowing well before dawn (ours have a penchant for 2 A.M., and/or whenever anyone in the house gets up for a midnight pee). It's the ewe lowing out in the field, calling for her misplaced lamb, and the livestock guard dog barking ferociously at the neighbor's taunting cat. It's the low clucking of dozens of chickens scratching through the grass and the soft grunt or shrill scream of a goat giving birth. And best of all, it's the scamper of tiny hooves across the barn floor each spring.

We'd timed it well: three goats due within days of one another, and all of them due less than one week before our big spring open-farm day. But there was just one problem. By the time Blossoms, Bees, and Barnyard Babies — the countywide open-farm event — rolled around, our third milk goat, Elizabeth, was still very much pregnant.

Which meant that, rather than having three mamas and their babies on display in the barn, we would have something that looked a little like a "before-and-after" segment. In one stall lay Elizabeth, her belly spread fatly across the straw, while in the neighboring stall frolicked Tuxedo, Sedona, and their four doelings.

On the day of the open farm, we started work at six in the morning. And it very nearly wasn't enough time.

Best of all is the scamper of tiny hooves across the barn floor.

All the little dangers we avoid daily — wire protruding from the fence here, slippery mud there — had to be taken care of before visitors arrived. We put down wood chips and fenced off places we didn't want visitors to go. We crafted signs that would welcome people, explain the "Things to Do" on the farm, and direct visitors toward the barn. We set up an animal table with goat, sheep, alpaca, and chicken feed for sale in little labeled Dixie cups. (This, we knew from experience, was far better than giving away food for free because there's always at least one child who will grab handfuls of food and fling it on the ground repeatedly. This way there's no wasted money, and less wasted feed, because parents are less likely to buy a second cup for a food thrower.)

Once the animal table was set up, we transformed our farmers'-market table into a display of information about our CSA program and pasture-raised eggs. We made yet another table for the sale of last season's winter squash. And then, the icing on the cake: We rearranged the hay bales in the barn to form stadium seating, as the space in front of the stalls would become a stage for a series of three workshops on goat care.

People, of course, started showing up twenty minutes before the opening time of 10 A.M. We didn't mind, though. Children and families poured into the barn and pasture, petting goats, asking questions, and writing their names and e-mail addresses onto a sheet we'd prepared.

Throughout the day, cameras went off like crazy, snapping photos of children holding tiny goat kids. Human mothers gazed at overfull Elizabeth with sympathy. ("I remember feeling like that" was the universal sentiment.)

At the appointed workshop times, the barn filled with people eager to learn about goat care. We distributed "grow-your-own-cheese" handouts we'd created, which gave fun facts about Nigerian Dwarf goats and their care. We went over the key components of a birthing kit, and how to tell when a goat was about to go into labor. And, naturally, we invited kids and adults alike to try their hand at eking a bit of goat milk out of an honest-to-goodness udder.

After cuddling with goats and tossing sunflower seeds to chickens, some visitors signed up on a waiting list for goat kids. Others signed up for our vegetable CSA, and still others purchased our winter squash, asked when the local farmers' market would be open, and promised to find us there.

Our adorable goat kids weren't just selling themselves; they were selling produce, too. It seems to me (and, in fact, to many of the people who visited) that a farm isn't a farm without this: the classic kids-with-kids moment, when a human child experiences a sudden rush of maternal affection toward another kid, one with four hooves and nibbling lips. In that moment, the human child realizes that the creature he's holding is not a stuffed toy, but instead a baby — a tiny thing, dear, helpless, needing protection. It's almost as miraculous as the sudden presence of another being in the room, that brief moment after birth when everyone becomes aware that what was once one is now two. (And then three.)

It's not just the children who are affected, either. Grown men and women sat transfixed by the baby creatures curled up asleep on their laps. It wasn't just children asking parents "Can we take him home?" but wives asking husbands and husbands asking wives. (We encouraged all the newly minted goat addicts to visit the farm again, and noted that we bring baby goats down to visit for on-farm pickup days at our CSA.)

Although animals are a lot of work — and hosting open-farm days are even *more* work — there is a payoff. A good farm nourishes not only the body, but also the soul. Our animals give customers the gift of the imagined farm from their childhood: the romantic ideal of a farm, one they can visit and enjoy without having to muck out stalls themselves. For us, this isn't just a neighborly or a poetic thing to do. It causes some customers to choose our CSA in the first place, and keeps other customers coming back year after year. And watching both adults and children become giddy in the presence of baby goats nourishes our souls as well.

This partly explains why we raise Nigerian Dwarf goats and Babydoll Southdown sheep. We chose these smaller farm animals for a reason: They're less intimidating than are their full-size counterparts, and they're breeds known for their friendly personalities. Our sheep and goats are kid-size, and backyard-farm-size, too; because of that, we can teach customers how to milk their own goats and make their own cheese, or how they can keep a couple of sheep to mow their lawn. Because our sheep and goats are small and thus easy to transport, we incorporate our animals into farm activities whenever possible: "chicken bingo" at farm parties, goat kids and a livestock guard puppy at the CSA pickup, and an open-farm day one week after kidding begins.

Visitors had been lingering, hoping that Liz would get down to business and have her babies during the open farm. But at five o'clock, we closed up shop. The last visitors made their way back to their cars. One woman, before leaving, gave us her phone number.

"If she goes into labor in the next couple of hours, would you mind calling me?" she asked. "My son really wants to see the newborns."

But it was not to be. Liz went into hard labor without dilating, and no matter what I tried, I couldn't fit more than one finger past her cervix. Soon I found myself driving eighty miles an hour down the highway with a laboring goat in the back of my station wagon as we rushed her to the Cotati Large Animal Hospital at ten o'clock on a Sunday night. By the time we got back home, got the kids to nurse, and ate a simple dinner in the barn, we'd been running around for eighteen hours straight. We hadn't eaten anything other than a hasty snack since six in the morning. But we had a story that we'd remember for a long time — one we'd tell to customers at the farmers' market, some of whom would approach us to ask about Elizabeth long after her wethers had been weaned and sold to new homes.

Even more important than the stories and the misadventures of life on a farm (which customers love to hear) is this simple message, given freely from a five-day-old goat kid to a five-year-old child: We're a farm; we're your neighbor. And all the little miracles of life that happen on a farm are here waiting for you, whenever you want them. ✍

Two Farmers, 350 Chickens, and a Hurricane

BY KRISTEN JOHANSON

Kristen Johanson lives and works on Blackberry Meadows Farm in western Pennsylvania, where she raises pastured poultry and helps run an organic-vegetable CSA. Four years ago she and her husband, Nate, made the decision to start a more sustainable life as farmers. They haven't looked back.

It was eight o'clock in the evening and I was doing my rounds in the brooder, caring for the baby chicks, while Nate was doing his rounds with the older broilers outside. I could hear the wind picking up and remember saying to the little chicks, "Be glad you're safe and cozy in here, little ones. It sounds wicked out there."

Just then, Nate came blowing into the brooder building. "It's crazy out there!" he said. "We're going to lose the covers on the broiler pens if we don't do something." So we ran out to see what we could do.

I had never experienced weather like that before. An eerie darkness had taken over the farm, painting everything a deep purple. Our broiler pens were made of two cattle panels, bowed over to make a hoop structure and covered with recycled billboard vinyl. The ends of the vinyl were loose so that we could roll up the sides during the day and roll them down at night. The wind was blowing so hard that the ends were flapping, threatening to pull off the vinyl and make it fly away. Every time the wind blew, the chickens, terrified of the flapping noise, would cower in the corner and pile on top of each other. We knew we had to do something fast or they would smother each other to death.

Nate acted quicker that I've ever seen him move before. He punched holes in the ends of the vinyl sides and tied them to the pens so they couldn't move. We had four pens of broilers, so we had to do

everything quickly. The wind howled and roared while Nate worked and I tried to weight down the sides with cinder blocks. Running out of blocks, I threw my body across the last one and waited for him to finish. The wind blew so hard that it even moved me.

Tears streamed down my face and I couldn't breathe. This was a huge blow.

We were in a pasture flanked by woods on two sides, and the trees were blowing around so fiercely that they were bending and creaking. I started imagining the worst, and I closed my eyes and said a prayer while Nate worked frantically. He finally finished, and the broilers seemed stable. It was well after ten o'clock when we accepted that we had done all we could and there was nothing to do but wait it out.

I didn't want to leave our animals — that was our livelihood sitting out in that field — but it was starting to become dangerous for us. Branches and debris were flying everywhere. We barely slept that night, waking with every howl of the wind. It broke my heart to think of the animals being so scared. We didn't know it yet, but we were being hit by hurricane Ike.

Morning came and we were afraid to walk outside.

This was our first year farming, and the learning curve was steeper than you can imagine. It was demanding, stressful, frustrating, exhausting, dirty, and beautiful all at the same time. When we took the leap into farming, overnight we became responsible for several hundred tiny little lives, and the weight of that responsibility was heavy. No matter what, our days were filled with hard work, and now the thought of anything being damaged and requiring more work was overwhelming. We couldn't afford a setback at this point in the game.

We hopped on the four-wheeler and drove over to the animals. Our first stop was the layers. We had recently moved our first batch of hens into the Eggmobile that Nate had finished building only a month before. We weren't prepared for what we were about to see.

In all the chaos of the night before, trying to save the broiler pens, neither of us thought to secure the Eggmobile. Ninety-mile-per-hour winds had lifted it up off the trailer it was on, rolled it 360 degrees, and crashed it down, right-side up, with all hundred of our girls inside. Hours and hours of Nate's work, smashed.

I thought he would lose it right then and there. Still on the four-wheeler, he turned to look at me, his face white. Tears streamed down my face and I couldn't breathe. This was a huge blow. We jumped off to check on the girls, hoping they were okay. Amazingly, every one of them had survived.

Once we knew they were all right, anger overcame Nate as he began to realize how much work lay ahead. And it wasn't like the work could wait. Those hens needed their home to be fixed so they could sleep in it that night.

We moved on to the broilers, to find that they too had survived the night, with minimal damage to the pens: a gift, a small bit of mercy, from the hurricane.

Nate and I looked at each other, looked over our battered farm, and breathed deep. There was nothing to do but rebuild.

We gathered our tools and began again. ✍

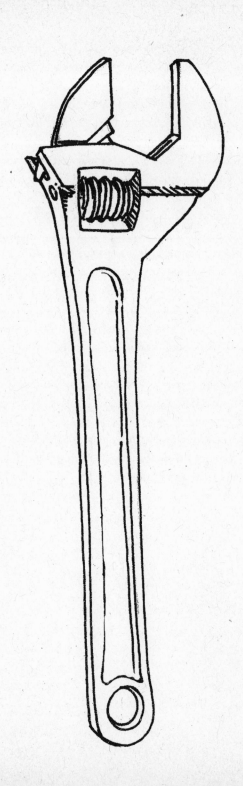

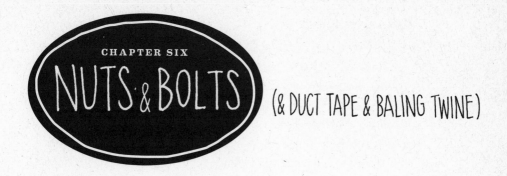

CHAPTER SIX

NUTS & BOLTS

(& DUCT TAPE & BALING TWINE)

Growing food usually requires a lot of stuff. Besides our bodies, land, and water, raising crops and animals involves tools and infrastructure: hand tools, power tools, heavy equipment, vehicles, barns, fences, greenhouses, irrigation systems, coolers. No wonder farming requires so much capital when we don't start off with all these things. What are we to do when we can't afford these useful items brand-new?

We make it work! We use our hands and noggins to retrofit what's available, keep on the lookout for used equipment, and create functional tools and fix-its with reusable sturdy things such as wooden pallets, cinder blocks, five-gallon buckets, and baling twine.

As for me, I feel most comfortable with the simple tools. I don't shy away from hand tools, and power tools are fun, but I'm still getting comfortable with our tractors. When my partner and I bought our first one and were thinking about getting another, a friend asked if we really needed two. We actually have three now. But talking to our mentor farmers, whose diverse farms are thriving, we think we're on the right track. The general idea is to use specific equipment for particular jobs that we can't do by hand.

My partner and I have chosen to farm full time, and being in California, with so many bigger organic farms to compete with, we've had to scale up vegetable production and start growing our own animal feed to stay financially viable. I hope we're staying true to the spirit of sustainability — creating fertility on the farm, preserving biodiversity, selling locally, and exploring the possibility of growing and making biodiesel for our tractors. Mechanization is not all about efficiency and the bottom line; we see it as keeping good farming and good food affordable and accessible.

For us, practicality came first in developing the physical basis of the farm. We couldn't afford a new tractor, so we settled on the Ford 5600 built in 1981: It had the right horsepower and we got a great deal. Farming in an ecologically responsible way involves good timing, and when we need to get something done, we git 'er done! Having good organization and the right tools, we're halfway there. That small window between rainstorms to cultivate a four-acre field would rapidly close if all we had was a rototiller.

With infrastructure planning, we can design for efficiency while making the farm beautiful. (Farmers are also artists.) In this chapter, the farmers illustrate their creativity and skill in making their tools work well to accomplish the job. There's a mix of luck and disaster in trialing a new farm-made tool or using new equipment, but safety and staying focused on the task pay off. After all the experimentation, going the DIY route can be fun as we evolve or adapt systems, buildings, and equipment with the aim of improvement — and profit.

Not everyone is a confident carpenter, mechanic, electrician, or welder, but our peers, mentors, and local resources can impart solid advice and teach new skills, and will sometimes even do it for us. If you look closely, there's a wonderful reflection of the farmer in the physical framework that he or she creates and uses to steward the land.

— Paula Manalo

On the Rise

BY SARAH HUCKA

With her brother, Joel, and the help of their parents, Sarah Hucka has been making a go of the true family farm in a quiet western Oregon valley for three years. They sell at three local farmers' markets and through an ultra-local CSA, where many members walk or bike to pick up their produce.

The solar-energy expert at the desk of the alternative-energy store didn't smile, and his eyes were half closed as he spoke.

"I can schedule you for a consultation next week. Then, standard procedure is to draw a design and give you a cost estimate," he said.

At least he was gracious enough to speak to my brother and me. When I first disrupted his stolid gaze, I feared he would be silent until coins were fed into a slot on the top of his head. This was the kind of alternative-energy store where you could compare bamboo to slate flooring while you have a solar hot-water system custom-designed. Never before had I had a reason to pass through its elegant doors.

"How much is the consultation?" I asked. Many of the pumps, panels, and inverters on display were worth more than my car. My brother's motto around the farm was "Cheap is good, free is better." Joel and I tried to do things ourselves, creating what we needed out of whatever the farm budget could afford. Occasionally we would spring for something complicated, high tech, or otherwise baffling. We were at the alternative-energy store to find out if our project was in or out of the "do-it-ourselves" category.

"My time is billed at forty-five dollars an hour," the energy expert said. At least it was less than I had feared. We could afford to hire him for a day or two if we absolutely had to.

"What's this over here?" My brother started walking toward the shelf where I'd seen a small pump. The expert didn't follow, so Joel picked up the display pump and walked back to the desk.

"That's an in-line pump. Eight-volt, DC."

"How much does it cost?" Our question signaled what the expert already knew: We weren't going to be big spenders.

"Two-hundred and twenty-nine dollars, and then you'll need the four square-foot solar panel that goes with it, and a deep cell battery."

Joel and I had researched how to heat our propagation greenhouse for better spring seed germination, and we thought we could use a pump to move warm water through hoses under the tables where the seedlings sit. Heating the bottoms of the trays is more efficient than heating the air, and cycling warm water seemed like a good way to distribute heat. Our farm is off the electric grid: thus my first visit to the alternative-energy store.

"Do you think this pump would move water through five hundred feet of hose?" I asked.

"I could sit down with you next week and figure out the numbers," the expert said.

I wasn't getting anything for free here. I gave a quick squinty smile and headed for the door.

"We'll call if we decide to go that route," I said, and walked outside.

Because that was the only alternative-energy store in the metropolis where we "go to town," the next natural stop was the thrift store. I have the good fortune to be the sister and farming partner of an experienced thrift-store shopper — my brother can identify the function of tools in the bottom of the deepest bins, where store employees have hidden things after being asked too many times, "What's this?" Sometimes Joel can't resist buying these gizmos, so we have a box in the barn with antique cherry-pitters, apple-peelers, and lever-action can-crushers. The hand-crank ice-shaver is a nice novelty, though I'm not sure about the clothes iron that's designed to contain an actual fire. Most of the farm tools and supplies in our shed are wonderfully useful and still bear their little red, yellow, or blue price tags for nostalgia.

We knew better than to rely on thrift stores for what we needed, but that day we needed a look around. Joel had already been at three of the four local thrift stores within the week, so we went to the one he'd missed. There weren't any batteries out back, no electrical wire on the shelves. No half-price solar-powered pumps or any other miracles.

Empty handed, we sat in his truck in the parking lot watching gray December raindrops smear the windshield.

"Well, I guess we could buy that pump," he said. "But we'd need a lot more than one little solar panel and one battery in order to run it all night, especially as we need the thing on winter days when the sun doesn't always shine," Joel thought out loud. "We could try to run it off the big panels we already have, but they're three hundred feet from where we need the pump."

My brother's motto around the farm was "Cheap is good, free is better."

We already run our well pump with solar power. Those panels probably had power to share, especially for a small, eight-volt pump. But we hadn't wired that system and didn't know the slightest thing about getting power from the panels all the way to the greenhouse. I knew from researching the option of connecting the farm to the grid that electricity dissipates along the length of any given wire. Three hundred feet was a challenging distance, never mind a lot of wire priced by the foot.

Where to go now? We didn't want to pay the consultation fee at the "green-means-dollars" store. That meant we'd have to turn this into a do-it-ourselves project and jump into the realm of the unknown. So, as much in search of inspiration as of information, we headed for the recycled-building-materials store.

The propane hot water heaters caught our attention on the open-air shelves. We'd crossed paths with steam radiators, but their world of elevated pressure and reinforced hose fittings was unfamiliar. Water seemed simpler. Now we just needed a way to circulate it through hoses under the seeding benches.

Joel stopped and turned to me. "Hot air rises, right?" he asked.

I didn't say anything, not sure where the idea was going.

"And so does hot water. So what if the force of that heated water moving to a high point would be enough to generate circulation through our hoses?"

Physics was the only class I'd gotten less than a B in throughout my college career. And I didn't get a C. It was a full-blown D.

I tried to give him an encouraging look. "Whatever you say."

I was the one who would sow the seeds, after Joel solved our heating problem. Neither of us could run a farm without the other — I was the green thumb, he was the greasy one. He built tools and fixed machines; I kept the books and went to market. This project was obviously under Joel's leadership.

"I think maybe if we get a short hot-water heater and slant our bench, this might work," he said. "But the concept is so simple, it's almost too good to be true."

One thing I can do is research. After we priced some parts at the recycled-building-materials store, I went straight to the Internet. I discovered the fancy name for the simple idea: thermo-siphoning. We learned that thermo-siphoning is a functional tool for heating things such as the floors of a home. But the detailed designs that I found relied on a cold-water source, such as a well or spring to feed the system. This source should have some force behind it, like a pressure tank feeding a water heater. We wanted our system to run by itself, with the same water cycling through the hoses on its own. I referenced this idea as a possibility, but I couldn't find any plans to follow that would ensure that our system would work. Joel would have to trust his belief that "hot water rises" and his love for tinkering to guide him through this project.

He started by laying out the hoses on the bench. Now, something warm needed to pass through them. He bought a length of copper tubing, coiled it into a funnel shape, and tried setting the wide end down on a propane burner. He hoped water inside it would be heated and rise through the coils, feeding a hose attached to the top. I don't know why it didn't work. I was a bystander, like the many neighbors who walk by our farm every day. We were all entertained by the clanging and muttering coming from the greenhouse, but we didn't ask questions. All I know is that the copper coil was soon sitting in the rain outside the propagation greenhouse while Joel tinkered with various tanks of water inside.

"I have no idea if this will work," he said one day, "but come help me straighten the hoses so we can try it."

On the ground below the propagation bench, he had a propane two-gallon water heater from an old RV. Various hose fittings came off the tank and fed five different hundred-foot garden hoses circling the surface of the bench. All the hoses eventually fed into a barrel of cold water, which, via a short hose connection, was the inlet back to the hot-water heater. We turned on the water heater and waited. After fifteen minutes, its thermostat clicked off the flame. The water inside had heated, but it wasn't moving through the hoses.

"Maybe we need to prime the system — make sure there's no air blocking the flow, and get it moving initially," Joel said. "Then it should propel itself. So, when I say so, turn on the spigot."

I opened the valve and waited. At Joel's yell, I turned if off again and rejoined him at the bench. Then we waited. The self-sufficient flow of hot water through the hoses was very slow. To know if the water was moving on its own, we had to wait a few hours and then check the temperature along the length of the hoses.

For the next five days, we monitored and experimented constantly, even checking the hoses a few times each night. We found that air bubbles in the hoses would stop the flow of hot water. We started a daily routine of flushing out air that accumulated, even during normal operation. We played with hose layout and length to achieve the least amount of resistance for the water. We adjusted the outlets from the hot-water tank, so that each hose had an equal opportunity for hot water to rise into it. And eventually, we were satisfied with the result.

We arranged the hoses one last time and covered them with perlite. We covered that with a piece of black plastic to keep the perlite dry. Finally I started seeding into trays. We achieved consistent soil temperatures of 65 to 70°F in the trays and air temperatures in the greenhouse of 50 to 55°F at night.

We're experts at it now. I'm thinking of billing other farmers when I explain the whole system to them. About forty-five dollars an hour sounds good. ℘

Potato Digger

BY ERIN BULLOCK

After growing up in the suburbs of Rochester, New York, Erin Bullock picked up the
sustainable-agriculture momentum living in the San Francisco Bay Area for six
years, then in 2009 returned home to start her own CSA, Mud Creek Farm. She now
leases twenty-eight acres of sandy loam in Victor, New York, grows fifty kinds
of vegetables, and is looking forward to acquiring land for a permanent farm soon.

The first year, we dug potatoes by hand. Eighteen
hundred pounds of them, with just one digging fork and lots of worn-
down gloves. We loved it. Each week we were supplying seventy-five
CSA members with vegetable shares from four acres of leased ground
in the suburbs of Rochester, New York.

I had just returned to the area after living in California for six
years, and I had a fresh career goal in mind that involved working my
butt off and getting very muddy. My parents, who lived fifteen minutes
away on the cul-de-sac where I grew up, were perplexed, but also glad
that I had chosen a location close to home. I had landed on a fertile
piece of ground holding tight to its agrarian roots, untouched by the
tentacles of sprawl.

One less-than-hectic fall day, I was walking through the "machinery graveyard" in the nearby woods with my city friends Doug and
Eli. We were trying to guess the uses for these rusty old antiques, most
of which were half sunk into the ground, falling apart, and indistin-
guishable by us young 'uns, new as we were to farming. Was that a
hay tedder? And what's a hay tedder, anyway? This was for planting
something . . . looks like the hoppers are missing. I wondered whether
Bob, the eighty-four-year-old landowner who had farmed these fields
for the last half century, had actually used any of this stuff, or if it was
left over from the previous farmers.

I had bought a shiny new Kubota tractor and borrowed a three-bottom plow from our fourth-generation farming neighbor, Jack. I was remembering that he had described this field we were working on as "old potato-growin' ground," and suddenly it dawned on me that this rusty old piece of equipment in front of me must be a potato digger! It looked about a hundred years old and had a hitch on the front to hook horses up to. We looked it over; most of the parts seemed intact, including its hard old iron seat.

Eli and Doug suggested we use our otherwise relaxed afternoon to try it out. There were no trees growing through it, so all we needed were a few shovels to free the wheels a bit and then strong young muscles to turn it around and push it fifteen feet to where it could be hooked up to the tractor.

We greased and oiled everything. There were a lot of working gears and levers, all functioning off the rotation of these big rusty metal wheels. I didn't have high hopes. I thought it would just fall apart before we even got to the field. We stuck a pin through the hitch into the drawbar of the Kubota and crept at the lowest speed out into the open. I have no idea when the last time this implement had seen a field. The sun made the oil gleam on the conveyor belts.

It creaked and limped . . . *click click click click click.* We all stared at the gears in wonder. It felt like some kind of revolution. Seated up on my orange diesel "horse," I valiantly approached a two-hundred-foot bed of potatoes, nervously watching behind for signs of disaster. We stopped at the front of the row, and Doug, sitting on the cold rusty seat, pushed the lever down so the plow point touched the ground. We drove forward. The plow plunged into the hill of potatoes and up came everything — the potatoes, the dirt, the rocks. The two sets of conveyor belts shook off the dirt as the potatoes rolled off the edge gently onto the soil. We dug the whole row. It sure was a lot less effort than digging on our hands and knees.

Looking at that creaky old tater-digger hooked up to the big new Kubota tractor, it was hard to deny the multigenerational imagery. Perhaps it was a metaphor for the cooperation between us and the veteran farmers we depended on for advice and wisdom.

We used the machine to dig all the potatoes our second year, when we doubled our production to feed a hundred and fifty CSA members. We're now going into our third year, doubling again to three hundred CSA members (eight solid acres of vegetables), at which point we'll cap our growth. I've seen bigger farms with fancy new potato diggers, but figure we'll try another year with Old Rusty. Diggers like this one were meant for small farms about our size anyway, back in the day when reliance on horses limited the size of their operations. I figure if it's lasted this long, it might have left enough spunk to keep going for another generation, if I take care of it.

That creaky old tater-digger was a metaphor for the cooperation between us and the veteran farmers we depended on.

Last year we got into some heavy soil that clogged up the conveyors and as we, frustrated, kicked off the dirt from the plow point, we thought we had broken it. After all the mud gave way, we realized that a few conveyor bars were just knocked out. They simply hooked back together, just like that. Old things really were built to last and to be fixed easily in the field. Thanking my elders, I started up the loud diesel engine again and finished off the row of potatoes. ✍

The Dibbler

BY JOSH VOLK

After working on other people's relatively giant, tractor-scale farms for a number of
years, Josh Volk started a tiny, part-time, hand-worked CSA farm two years ago. In
his "free" time he tries to help other folks with their farms by writing, consulting, and
designing new tools.

For six years, every time our crew headed out to the
field to plant, they would haul a modified fifty-five-gallon drum with
them. They'd pull it by its black-pipe handle down the beds, so that the
stubs of pipe welded around the drum could dibble marks showing
where to plant seedlings. That drum, dubbed "the Dibbler," had probably been pulled down a hundred miles of beds and farm roads. The
handle was attached to axles with iron plumbing Ts that creaked and
groaned rhythmically. Everyone on the farm recognized that tune and
its variations with the weather. Even after someone drove the old Ford
100 over the handle a few years later, putting a nice little bend in one
side, it kept creaking away, up and down the beds.

When I first built the thing, I didn't intend for anyone to have to
pull it (even though one farmworker, Nate, spent so much time pulling
it that he became one with it, and everyone called *him* the Dibbler).
What I had envisioned was something that would hook to the back
of our spader behind the tractor. My idea was that while the spader
formed a nice fluffy bed, the Dibbler would trail behind, leaving marks
for the planting crew. The handle had two extra pipes sticking out of
the sides, which I'm pretty sure no one else on the farm ever understood were intended to line up the Dibbler with the side plates of the
spader. I'm pretty sure of this because after a few attempts at running
it behind the spader, I gave up and realized it would work better if
someone just pulled the thing after I marked the bed pathways with
the cultivating tractor. The five-foot piece of all thread rod that I

bought to hold the thing on to the spading machine ended up in the bin of spare metal parts: too useful to get rid of, too obscure to look like anything other than junk to the untrained eye.

Among the other pieces of useful junk that accumulated on the farm over an impressively short period were several lengths of bent steel pipe. These were long sections that at one time had been part of a high tunnel that collapsed under a half-inch sheet of ice (another story for another time). Anyway, with the partial success of the Dibbler as a useful farm-built tool and a drip winder (also made from black pipe), I saw one of these pieces of steel pipe as an excellent candidate for building a mechanism for rolling out, and rolling up, floating row covers, which we were using miles of. What I needed were two big wheels; the pipe could be the axle.

I located a wire spool, the kind the telephone company uses to run out massive lengths of cable. The spool was perfect, with four-foot-wide rounds that I was convinced would make great wheels. Four skinny, three-feet-long bolts kept the center of the spool sandwiched between the two rounds, so I removed them, separated the rounds, and made two wheels and a small pile of scrap lumber. I was sure those bolts would come in handy at some point, so I tossed those in the bin of spare metal parts and took out the straightest piece of high tunnel pipe for the axle.

Unfortunately, the resulting row-cover winder was a bit of a disaster. The rounds were way too heavy and the whole thing was so awkward that even though rolling up the row cover by hand was one of my least favorite jobs — frequently a cold, wet, muddy, forearm-burning exercise — I temporarily admitted defeat.

Eventually my old projects come back around, though, and new ideas on how to implement them start to form. Six years after my failure to mount the Dibbler on the spader, my process for creating beds had gotten even more complicated. Instead of making one pass to form and mark a bed with the spader, I was now making three separate passes: one with the spader, one with the cultivating tractor with a gang of three Planet Jr. seeders to mark the lines, and one by someone pulling the Dibbler to mark the in-row spacing for the plants. Although all three passes were doing something beneficial, more passes by the tractor is not better when each one costs time, diesel, and soil compaction.

Actually, the bed-preparation process was even more complicated and imperfect than just needing three passes instead of one. For transplanted crops, I liked to mark the lines with our cultivating rake instead of the Planet Jr. seeders. This kept the soil loose, easier to plant into, and less likely to germinate weed seeds. The problem was that for all six years I had trouble marking the lines clearly enough with the rake. Even when the lines were clear, there was always some rogue on the crew who didn't understand exactly where I wanted the plants in relation to the lines. Additionally, if the soil conditions weren't just right, it could be difficult to see the lines, not to mention the Dibbler's dibble marks, as they were just small indents between two faint lines.

Nate spent so much time pulling the drum that everyone called *him* the Dibbler.

For direct-seeded crops, I would take off the rake and mount an entirely different tool on the tractor. I liked to use a gang of three Planet Jr. seeders to mark the lines because they left a nice clear firm pathway for the walk-behind Earthway seeder, and that firm pathway seemed to help improve germination. I know it sounds silly that I was marking with one seeder and then sowing with a different seeder, but believe me, it worked better and even faster that way. In fact, it worked so much better that I felt I could justify the extra work and effort of switching out the rake on the tractor for the seeder gang every time I wanted to mark the beds. I desperately wanted to reduce that complication and added workload, though.

One day, touring a friend's farm, I noticed a big gang of partially rusted-out Planet Jr. seeders in his metal spare parts pile (which was being consumed by weeds, as is commonly the case on farms — at least the ones I visit). He wasn't using the seeders, so in exchange for some help setting up a cultivator for his tractor, I took three of the press wheels off his hands. My thought was that I didn't need the entire seeder assembly just to be marking rows, and that I should build something lighter and easier to put on and take off the tractor. I also wanted to avoid wearing out our own perfectly good but somewhat aged Planet Jr. seeders. The circumference of the press wheels was just about three feet. What I needed was a long axle that I could put all three wheels on at their proper row spacing and then three rods to attach to the wheels, to mark each foot of rotation.

I don't know of any shop where you can buy a skinny bolt that's three feet long, but as luck would have it, I had four of them left over from that spool I'd taken apart four or five years before. Fortunately, my partner's threat to take all of the "junk" metal to the recycler earlier that spring hadn't come true yet, and searching through her consolidation of my metal piles, I came across the four bolts. Unbelievably, the bolt was the exact size I needed to slip through the bushings on the wheels. With a few spacers made from a bit of half-inch conduit, the other three bolts cut down as the rods, too many clamps bent from some failed cultivator project six years earlier, and pieces of an old bent-up truck rack I'd built for carrying seedlings years before when our greenhouse had been three miles down the road, I had cobbled together a new, improved Dibbler in just an afternoon.

This Dibbler was silent, mounted easily on the tractor, and marked both for beds to be seeded with the Earthway and for beds to be transplanted, putting down a clear wide mark every foot. So there I had it — a slightly less complicated way of marking beds and fewer passes to make over the field. In addition, the marks made by the new Dibbler were much clearer than those from the old one, thus easier to follow by a fickle crew.

I still haven't come up with a good way of picking up row cover. There are still piles of old metal around the farm, though, and I'm always thinking about how I might one day put them to good use. ✀

Tackling a Beast

BY ADAM GASKA

Adam Gaska grew up in Redwood Valley, California, and owns Mendocino Organics, a diverse biodynamic farm. He enjoys farming vegetables, grain, feed, and hay with his tractors.

I find myself once again staring at a beast covered in dust and grease. No, I'm not looking in a mirror. I've got my hands in the belly of our beast, a John Deere 4020. Lucky for us, there are just a few old, cracking hydraulic hoses in need of replacement. It's going to set us back only a couple of hours of wrenching, straining, and cussing to find the problem, and a hundred dollars to solve the problem. That's a pretty cheap way to appease the diesel-burning gods these days.

Growing up, I never thought to be a mechanic or a farmer. Sometimes things just work out the way they want, personal thoughts in the matter be damned. In high school, what I really wanted to be was an engineer, and my academic path was geared accordingly. The problem was, my class schedule easily filled up with math and science with room to spare. Not really being a jock or an artsy or performing type, my options were pretty limited, so I settled on auto shop. I figured, what the heck, it's closer to math class than choir is, and it'll help me later in life. I'll be able to save a few bucks by making minor repairs on my own car or at least know enough not to get ripped off by a shady mechanic. Throughout my high school career, there always seemed to be room to take another vocational class, like welding.

Once I came to the realization that I wanted to farm, those classes actually paid off. Not only did I learn to fix a few things, but I also felt comfortable hanging out with and learning from the older, chaw-spitting farmers who became my impromptu mentors. Sometimes the best gifts don't come in little boxes with nice little bows; sometimes they wear dirty coveralls and say things that'd make your grandma blush.

There are plenty of places to learn how to fix farm equipment aside from seasoned vets (although that's really where it's at). Many community colleges have vocational-education courses in mechanics and welding. The local tractor dealers, where I've found myself more than once, are a valuable resource. Generally, the mechanics there don't mind answering some of my questions.

Besides my roundabout education, I'd say my next best investment has been in tools. Whether it's a mechanic's bill or an invoice for a good ratchet set, it can be written off on my taxes. A good set of tools, along with a bit of education, will serve me for a lifetime. Trust me — if you plan on farming for your lifetime, you'll need both.

I get the hoses put in place and tightened up. I turn the old gal over a few times, and vroom, black particles sprinkle the air. I inspect the job and I'm pleased; no more leaks, and neither the patient nor the doctor is the worse for it. Although I usually don't walk away from any task looking clean and tidy, today my tools and my experience have served me well.

Don't get me wrong. I don't fix everything. Usually I know when something is beyond my skills and that I must enlist the help of someone with much stronger magic than I have. Fortunately, by keeping up with routine maintenance and dealing with small problems before they become bigger (and more expensive), that doesn't happen often.

Today I'm victorious, and I ride my iron steed, breathing diesel smoke and not leaking hydraulic fluid, off into our next season. ✍

A good set of tools, along with a bit of education, will serve me for a lifetime. Trust me — if you plan on farming for your lifetime, you'll need both.

The Right Tool for the Job

By Brad Halm

A small garden at the Homestead, an experiential living option at Denison University, was the spark for Brad Halm's interest in growing food. After working on organic farms around the Midwest, he moved to Washington state to help Colin McCrate start the Seattle Urban Farm Company. He's been building urban farms ever since.

Tools are now and have always been vital to farming: They're how we interact with the land to get things done. Whether it's a stirrup hoe or a cultivating tractor, our favorite tools become extensions of our bodies, as we use them again and again.

We farmers have a very interesting relationship with our tools. Nowhere have I seen more random and unlikely materials become useful tools than on a farm. As a group, farmers seem to be driven by the ethics of rugged individualism, frugality, and independence, and there's no way in hell we're going to pay $59.99 for something we can make for free from scavenged materials from around the farm (even if it takes us two weeks to make it).

Small farmers are also caught in a void. "Gardening" tools sold at nurseries are largely overpriced, poorly designed, and way too dainty to get anything done on a farm. Conversely, most farm equipment is sized for huge producers with thousands of acres of commodity crops and is obviously unsuited for us smaller-scale folks. Thus, we're forced to repurpose odd items found hidden in the recesses in the back of the barn, to buy antiquated equipment at auction, and to improvise our own tools.

For me, there was no better introduction to this phenomenon than apprenticing for a few seasons with Roy Brubaker on his organic vegetable farm in central Pennsylvania. Roy had a fierce imagination and a steady hand with an arc welder, and he hated to buy anything. This combination led to a steady stream of inventions pouring from

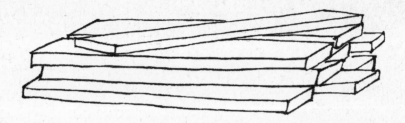

his shop, some absurd and some amazing. I remember harvesting greens with a tool made from a food-processor blade and an electric drill, "harvesting" pesky sparrows and invasive starlings in a human-size trap, and watering transplants using harvested rainwater from a reclaimed milk tank on a semi. We even had an old Dodge Colt that was converted for use as sort of a field-delivery wagon. We'd drive it down the farm lane, filled with sacks of fertilizer for the transplanting crew, and return to the packing shed overloaded with boxes of melons or bunches of beets from the harvest crew. Once, in a pinch, I picked up my mother from the hospital after a serious bacterial infection in our "Crop Car." It was actually a pleasant ride. Mom napped in the front while a fresh breeze whipped through the missing window seals and wafted the scent of organic fertilizer through the car.

Roy loved working in his shop and would do so as often as the constraints of farming would allow him to. I remember him struggling through a cold spring day of harvest with a serious cold, hacking and coughing all the while. Later that night, I found him in the shop, the temperature even colder, welding away. "I'm starving my cold," he said. I think he figured if the conditions were tough on him, they had to be even worse for whatever virus had infected his lungs. Sure enough, the following day yielded a chipper and smiling Roy, complete with a new creation from his shop in hand.

He never passed up the chance to create his own solution for whatever problem arose on the farm, even if he could buy one relatively inexpensively. We had homemade opening devices for our greenhouse ventilation windows, a homemade spool to distribute our plastic packing bags, and a homemade barrel washer to clean our root crops.

Were all these homemade tools worth the effort it took to make them? Did they really improve our efficiency? Did Roy get a proper return on all the time he spent in the shop? The answers are immaterial. I learned a lot from working with someone who was unafraid to exercise creativity to solve a problem rather than reach for an off-the-shelf solution.

During my second winter in Pennsylvania, I was midway through building a set of harvest crates from a poplar harvested from the farm when I realized that, for me, the most valuable part of farming lies in the process, not in the number of pounds of produce harvested at the end of the season.

Small-scale farmers are forced to repurpose odd items, buy antiquated equipment, and to improvise our own tools.

That said, tools for small farmers are starting to become more available. New companies are springing up and old ones are expanding their product lines to supply high-quality equipment to farmers of all scales. It's a welcome change, as there's nothing more frustrating than struggling through a task with a poorly designed tool. Ultimately, I think tools are at their best when they're well planned, properly made, and efficient to use but still carry some of our collective farmers' soul. My favorite chore is digging burdock roots using the "Kentifer" weeding tool that Roy made for me in his shop from a piece of old truck spring steel (the original model was built as a wedding gift for two of his earliest apprentices, Kent and Jennifer). It's strong as hell and a joy to use, and brings back a lot of good memories every time I pull it out of the toolshed. ✄

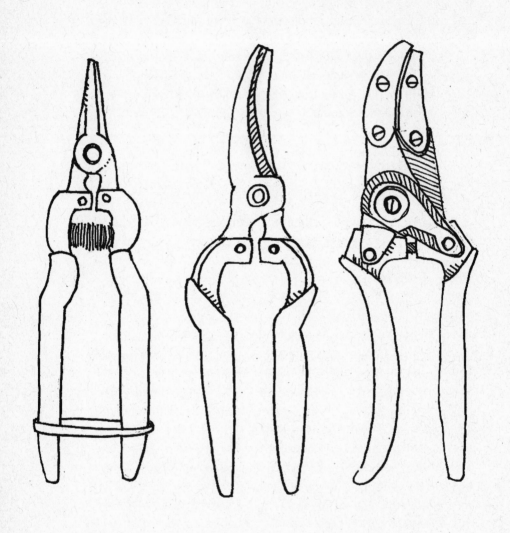

CHAPTER SEVEN

NINJA TACTICS

Perhaps by now we've convinced you that agriculture can be quite a complex negotiation. And that this, it turns out, is part of the thrill. Doing something hard, particularly when times are tough: that's when the heroics of a difficult path seem most attractive. It may not be easy, but it sure isn't boring. If we had been born in a different era, some of us might have become pirates or whalers, beaver trappers or Klondikers; we might have been mercenaries or missionaries, exploring botanists, or sod busters. Pioneer spirit doesn't capture it completely (although that's certainly part of it). What we're talking about here are the ninja tactics of starting a farm, and a farming career, despite the odds.

Driven by spring-loaded opportunism, we're ready to make things happen. We take a mental inventory of the world, always scouting for useful objects, fixtures, and components that can be retooled. In the city, it means Dumpster diving for pig slop and building supplies. In the country, it means joining town boards, getting hooked up with cardboard from the appliance store and sawdust from the lumber mill. It means surfing Craigslist for shop tools, hitting up yard sales, fact-finding at the county offices about barns that look vacant, finding benevolent landowners and storage-space landlords. A lot of bobbing and weaving and institutional interface.

It means giving back, walking a straight path, being accountable in a serious way.

Establishing the radical farm goal is one thing; making nice with all the parties along the way to that goal is another. It doesn't take long to figure out that nimble maneuvering, staying positive, and strategic generosity go a long way to ensure goodwill in the community. Big things start small.

We're ninjalike because of our ambition and struggle to stay aloft, because the ordered space we're working to build must compete at every junction with disorder. We're about discipline in a world of convenience, about cooperation and reciprocity in a world in which the idea of the individual is pervasive.

So yes, agriculture is hard, it's dangerous, it's a big commitment, but no, it's not boring at all.

— Severine von Tscharner Fleming

Apprentice Tent

BY JEN GRIFFITH

After working on agricultural projects in Nicaragua, California, Tanzania, and New York City, Jen Griffith is currently an apprentice at Quail Hill Farm, on the South Fork of Long Island, where she is close to two of her loves — the Atlantic Ocean and Brooklyn.

I boarded a plane from New York City to San Francisco, checking my whole life, housing and all, in the cargo compartment beneath the plane. My destination was an organic farm where apprentices spend the season in tents.

Coming from a city where space is a precious resource, I bought the largest tent I could find — a twelve-by-nine-foot canvas army tent tall enough to stand inside. Once I arrived in California, I scoured Craigslist for a queen-size futon, a dresser, a bookshelf, and carpet. Using graph paper, I created scaled diagrams of how I would lay out my new home.

I erected the tent on the north side of the farm, with the front door facing a long, thin field of nondescript plants. The farm was a patchwork of greens, textures, and long rows plowed in all directions.

Going to bed that first night, I bundled myself into sweat pants, put on a wool hat, and wrapped myself in every blanket I had. California nights were colder and damper than any East Coaster might imagine. Inland from the Monterey Bay, it gets very windy, as the warm ocean air rushes in, as the land air cools at night. In the middle of that first night, I woke up to find my tent shaking. The winds ripped and pushed against it, bending metal poles and flapping fabric.

I didn't sleep much at first, as I adjusted to the sound of canvas rippling around me. I felt so exposed to the elements.

Some nights I fell asleep to the hooting of the barn owls in a cypress tree above me. Most nights, I could feel the fog from the Pacific Ocean roll in and leave a coat of dew on everything, including me.

Living in my tent, the fields became my hallways. I'd unzip the mesh tent door, hair messy and pillow lines in my cheeks, to arrive at the farm. Running on caffeine and the adrenaline of a new experience, I worked hard. I struggled to tell the differences among chard, kale, and collard. I fumbled trying to figure out how to hold a hoe so that my back would ache a little less. On foggy days, my pant legs were wet at the bottom; on dry days, they were dusty.

Eventually, the physical exhaustion set in, and I was able to sleep through most nights.

Intermittently throughout the season, I would still wake up to a pack of coyotes laughing hysterically after they made a kill in the field just behind my tent. At these times, the blood would rush out of my body and the feeling of exposure would return. This thin canvas wall was all that separated me from an active nighttime world.

As dirt rubbed into the cracks of my hands, the farm started to seep inside of me. Change is funny. Sometimes it happens in small, imperceptible increments; at some point, they add up to something you can see. The plants outside my front door became a small field of wheat. Every day I watched the plants progress from tiny, bushy leaves to sturdy, upright stalks to flowering green heads to golden grain, ripe and ready. We harvested the wheat right in front of my tent, cutting it with sickles, threshing it against pallets and tarps, and winnowing it in the Monterey Bay breeze.

Throughout the summer, I spent less and less time in my tent. It had started to seem silly to me how much I fixated on it when moving out there. I now spent free moments in the strawberry patch, gorging on berries, or under the kiwi arbor for a private afternoon nap. One evening, after work and before bed, I sat in the apple orchard to watch the sunset. A great blue heron swooped down and landed twenty feet from me. For fifteen minutes we sat together quietly, side by side. Suddenly, it thrust its head into the ground and came up with a gopher in its bill. It tilted its head back, tossed the gopher, and swallowed it whole. I could see the body working its way down the heron's skinny neck. That night, I went to bed tingling. ✄

My First Intern Worked Pregnant for the Entire Summer

BY ERIN BULLOCK

After growing up in the suburbs of Rochester, New York, Erin Bullock picked up
the sustainable-agriculture momentum living in the San Francisco Bay Area for six
years. She then returned home to start her own CSA, Mud Creek Farm, in 2009.
She now leases twenty-eight acres of sandy loam in Victor, New York, grows fifty kinds
of vegetables, and is looking forward to acquiring permanent land soon.

My first intern worked pregnant for the entire summer.
Beth had walked into my life at exactly the right time, moving back to
her hometown, just like me, after ten years of living away from it. She had
just gotten married, and she and her husband were looking for a house to
buy. She wanted to learn how to farm. She had volunteered with us for a
few days in the fall, so I knew she was a hard worker and already knew a
bit about growing food. We picked out seed varieties together in January.
We even discussed having her be a partner in the business. Eventually,
we settled on an employee-employer relationship, forty hours a week at
minimum wage, April through October. I had her draft a "learning contract," so we could check in periodically to see how we were both doing.

I was nervous going into it. Would I be able to balance the responsibility of mentoring with the realities of production agriculture?
We would have a hundred and fifty CSA members that year. I really
wanted her to have as good a mentor as I had when I was learning
how to farm. Dave Hambleton, at Sisters Hill Farm in Stanfordville,
New York, had walked me step by step through how to do everything
from thinning beets to using a moldboard plow, and inspired me to
start my own farm. Here I was, only two years later, on the other side
of that learning relationship. I was scared.

After a few weeks of working together, I expressed my concerns (we tried to keep open communication with each other about everything) and Beth replied, "Don't worry about it. I'll pick it up as we go. Just focus on getting things done, and I'll ask questions if I need clarification."

Gosh, what a relief.

We were a great team. She was a people person, and helped motivate volunteers in the fields. I learned how to designate tasks, when to train her in something and then let go. I remember a big cultivating day in April, when I taught her to drive the Farmall Cub with the cultivating sweeps. I watched her go up and down the rows a few times, hitting fewer cabbage plants than I usually do (it was beautiful), and I walked away from that field to do something else. It was hard, because cultivating is something I enjoy doing. At the same time, though, things were getting done, and she was learning a new skill.

One day I took the plunge and trained Beth on the big tractor (Big Orange), my shiny new fifty-horsepower, four-wheel-drive Kubota — a twenty-three-thousand-dollar investment, worth way more than all of my other equipment combined. But it was easy to drive. What took a little while to learn was how to use the front loader, so for a morning I had her turning the compost pile and moving wood chips. Meanwhile, I was working on other things, getting kind of frustrated at how long she was taking for these simple tasks that would have taken me an hour. I went over to see how she was doing and noticed that her face contained even more frustration than I felt; she was even cursing. I was thinking, "Uh-oh. Here's a side of her that might make things difficult this summer."

We sat down at break and talked it over; Beth was very emotional, and I was somewhat concerned that she was questioning her decision to be working here. I decided to give her a rest from Big Orange and have her do some weeding for the rest of the afternoon.

The next day she seemed calmer but still a bit reserved, like she was holding something inside. At lunch, I asked her if there was anything going on. She said, well, she might as well tell me: She was pregnant. She thought I would be angry. (I was.) She told me she still wanted to work until the end of her contract, at which point she would be almost eight months along.

"If lots of women have done this before," she said, "and still do in many other countries, I can, too."

It took her a long time to start looking pregnant. She would occasionally have to sit down or stop and stretch when a muscle cramp caught her unexpectedly, but otherwise she could do everything I did: squat and weed a row, wield a hoe or a pitchfork, harvest, haul sandbags, move row cover, fix irrigation, lift sacks of cover-crop seed, drive the tractors. She was strong. As a matter of fact in September, when the harvest got heavier, we had to yell at her to not pick up the full bins.

Volunteers pitched in to do the lifting. She had a hard time with it; she didn't like feeling weak or helpless.

Meanwhile, her husband had found the perfect farmhouse — an old fixer-upper with three acres in a nearby town. They were doing renovations and working out the details to transfer ownership. I went over just before the farmhouse was scheduled for its final inspection, and we had a little picnic on the front porch. There was a big maple tree in the yard, and I had this clear image in my head of her pushing a child on a swing under its branches. This would be their home, their nest, and I shared in a bit of the happiness of what that meant.

The next week it burned down — right down to the ground. Beth's husband was driving there to work on it and saw the fire trucks. The house was leveled, and all his tools were lost. They never found out the cause, but it was an old house, so it was probably some kind of electrical fault. At least no one was inside when it happened.

The couple believe in signs. Beth showed up the next day to work, and it was a quiet, solemn day for us all. As devastating as it was, I'll never forget how quickly she regained strength and moved on, always looking forward, making do with what life dealt her.

The final melon harvest was in late September. Beth and I were in the field, knocking on watermelons, listening for that hollow thump, and tossing them to folks who would toss them to other folks who would place them on the wagon at the edge of the patch. After an hour, we took a break. Beth said it was enough melon-tossing for her.

It had taken her awhile to say it; you could tell it pained her not to be part of the action.

For the month of October, she phased down to thirty hours a week, and I worked a bit extra to cover for her, but things were slowing anyway. It was all harvest now, and she was helping to manage crews of volunteers, whom I had already lectured about being proactive about picking up heavy things. During the last few weeks, she couldn't climb onto the tractors very easily, and fit into just the XXL rain paints. She did a lot of the veggie-washing at the tubs, as even bending over to harvest was difficult.

We had a lot of laughs at lunchtime, talked a lot about midwives and doulas, and cursed hospital C-sections. Everyone had questions for her: What did it feel like when the baby kicked? How much does it squish your bladder? Finally, on October 31, the last CSA distribution day, Beth was ready to stop working. As strong as she was, she said she felt like she couldn't bear another day. She needed a month of rest and preparation for the birth.

For the entire week before she was due, I couldn't sleep without dreaming about the birth. Then I got word that she'd delivered a beautiful girl, Emilia. She gave birth in her mother's house, without any painkillers. The next week, I talked with her, and she described the experience in detail. After three days of not sleeping or eating, as her contractions got more intense, they were ready to send her to the hospital if she didn't have the baby before morning. She told me that threat gave her the last bit of motivation she needed to push. She was crouching, then standing up to rest, crouching and pushing more, and finally Emilia was born. The midwife said she had never seen anyone with such amazing crouching muscles before, and Beth credited it to a season of serious farmwork.

Now I wait for spring to thaw the ground again, and the sunny days of next season to come. Meanwhile, I imagine a little toddler running through the sparkling blue-green rows of cabbages, as if she already knows a thing or two about cultivating. ✿

Farming in a Climate of Change

BY GINGER SALKOWSKI

Ginger Salkowski and her friend Brian Schulz own R-evolution Gardens, an off-the-grid farm and educational center in the Nehalem Valley, on the northern coast of Oregon. The farm is devoted to helping people make the journey toward self-sufficient, sustainable living.

"Is it time to freak out yet?" I ask myself, staring at the handful of crumpled bills that represent what's left of my savings from the previous year.

It's January, and that's a lean month for farmers. The CSA money hasn't come in yet, although I posted my annual call for new members a month earlier than I usually do. The farmers'-market money that I managed to parcel out of an envelope in my desk drawer over the last few months — for mortgage payments, gas, property taxes, phone, and Internet — is now gone. I've committed to taking two more acres under lease this growing season, and in my head I add up how much deer fencing, soil amendments, drip irrigation, compost, row cover, and seeds will cost this spring. The total of my projected expenses doesn't inspire flat panic, but maybe it should, considering that my current investment capital is stuffed in my back pocket.

The way I look at it, in year three of my farming career I'm in basic training boot camp for the future. I'm a single, mid-thirties female with no money, no fossil-fueled machinery, no real training or farming background, growing food for a fifty-family CSA, selling at two local farmers' markets, and holding educational classes on my off-the-grid land. That's called jumping into the deep end of the pool and hoping you learn to swim really fast.

Everything about my entry into farming has been fast and unpredictable, yet nothing I've ever done before has felt so right and so urgently important. My working theory is: Being a crazy upstart farmer may be the best hope I have, as the climate begins to change on planet Earth.

There's no good news about climate change. Nary a reputable scientist is left who doesn't agree that we're entering into a phase in which Mother Earth has some serious mood swings. Climate models show effects on a scale from devastating to catastrophic as the human love affair with fossil fuels over the last century begins to unravel. I don't intend to convince you of the hard science. There are hundreds of books out there with all the data you need, if you can face it. I'm writing this because I'm a vegetable farmer, and anything that has to do with a change in the weather has my complete attention. The current global forecast is pretty scary stuff and something that anyone entering into the field (pun intended) needs to study up on. That said, this farmer has learned better than to argue with the weather. The question I ask now is: What can I do about it?

My first summer farming, I was tending an acre of diverse organic vegetables pretty much on my own. Every day was a whirlwind learning curve. I often woke at six in the morning and worked until after nightfall, standing alone in the field in the dark with watering hose in hand, wetting down lettuce transplants as my housemates drank beer and laughed around summer barbecue fires.

Being a crazy upstart farmer may be the best hope I have, as the climate begins to change on planet Earth.

One day my chickens got out of their run and into my garden. I came back from a delivery and found a flock of forty hens decimating a field of seedlings that I had painstakingly transplanted by hand just a week earlier. Already exhausted from a day of hard work, I found myself running around like a madwoman, yelling at chickens and discovering one ruined bed after another. Luckily, a friend showed up in time and we managed to lure them back into their run with a scattering of fresh feed. Once the hens were secured, I slumped next to a fifty-foot, newly planted bed of chard. Everywhere the remains of what were healthy little plants were scattered and uprooted. It was so painful to look at that I burst into tears.

I spent the next three hours crying and repairing with my friend's help what we could of the damage. That summer, I vowed I would become a successful farmer if it killed me. It almost did, but I'm finally getting closer to understanding the meaning and terms of that success.

A successful new farmer in today's (and tomorrow's) climate has to have a serious package of skills. You have to be able to live with less. That means you have to be able to get extremely creative with very little money and time in order to make your season happen. You have to learn to thrive on uncertainty, seeing each new problem as a challenge that will make you a better farmer in the future. You need to be strong in body, especially in the lower back; stretching and yoga become crucial. You must be strong in mind: Can you figure out how many broccoli plants a fifty-family CSA will consume each season, and how to keep the succession plantings in rotation with thirty other crops in a small field, and, from that, know how many seeds you'll need? And you must be strong in spirit: In times of high stress, there is grace to be found in pausing to observe the first sweet-pea blossom opening in the morning light, in savoring the taste of a carrot you sowed as a tiny seed. Farming is a spiritual act; it's life, death, and rebirth every day, played out in millions of little scenes across the land. In this job, I feel lucky to be there to witness it all.

As I develop the skills of a wily farmer on my five acres of coastal clay, I'm becoming a new person altogether. I'm now someone with the flexibility it will take to weather the coming storm of climate changes. I'm someone who's no longer afraid to look down the barrel of a growing season with no cash, because I trust in the community of support that I've built here in my little seaside villages. These folks have helped my farm grow from day one and are invested in my survival to meet their food needs.

I fear no power outages brought on by increasingly violent windstorms, because the rainwater in my creek turns a micro hydro-turbine that keeps our farm lit up in winter and solar panels keep our energy and hot water flowing in summer. Firewood cut from our own forest gives us heat. I'm someone who can grow lots of different things and in many seasons. Unlike the vast cornfields and wheat fields back home in the Midwest, my farm maximizes diversity. I try to grow everything I can in this climate and use hoop houses and low tunnels to extend the season.

Perennials such as asparagus, raspberries, and sunchokes are in the mix to make resilience to climate and weather stress even stronger. I study permaculture, biodynamics, and French intensive farming, and I try them all. Whatever produces food and makes my soil healthier, I keep; things that sound groovy in a book but drive me nuts in real life (like wheel-shaped beds) I try once and then move on.

Because of climate changes, I may have to leave my farm one day and seek out new growing territory. Farms that were once productive may become too arid, too flooded, too cold. I need a set of skills (and seeds) I can put into my backpack and move to another spot to start again. We need to face that this may be the future of farming. There is no better way to get ready than to jump in and start learning. The scrappier and more resourceful you are, the greater the degree of success you'll have in the long run.

And remember that part earlier about it also being the most rewarding thing I've ever done? It is. I love every minute of my crazy job. One of the ironic results of the climate-change crisis may be that we humans will finally understand that we and our planet are interdependent, and we need to nurture that relationship. Only then will we discover that this connection is our deepest source of happiness and inner peace. ✄

Weathering the Storms

BY KATIE KULLA

In addition to helping run a vegetable farm with her husband and their son,
Katie Kulla writes about farming for the CSA newsletter and for journals such as
In Good Tilth and *Growing for Market*. Their farm is called Oakhill Organics,
in Dayton, Oregon.

SPRING 2006

My husband, Casey, and I are constructing our first
greenhouse. We moved to Yamhill County, Oregon, just a few weeks
ago to start our own farm. Even though we worked for another farmer
in a different state for two years, we don't know what we're doing.
We're struggling to put up a greenhouse that's ultimately much too
tall and weak for the windy site we now call our farm home. We're on
a time crunch. It's already the end of March, and our first market is
June 1.

Although the spring rain keeps falling, we have to build our
greenhouse so we can start our first seeds for transplants. We work
through the weather, erecting steel greenhouse legs with thunder and
lightning in the background, and pulling the poly skin over our two-
story-high greenhouse during the calm early morning before the
day's wind begins.

As the winds pick up, I watch my husband tie a rope around a
bunch of heavy, loose poly at the open end of the greenhouse, struggling to secure the plastic to the ground. The strong gusts keep lifting
the poly twenty-feet into the air and snapping it loudly. My husband kneels on the ground, attempting to hold the rope with all his
strength. And that's when I realize it: We're going to lose. This wind
is much more powerful than we are. And I want my husband to live.

"Casey!" I yell as loud as I can. "It's not worth it! Let it go!"

Eventually, as the wind dies down a bit, we manage to tie the plastic in place temporarily. After dark, I change out of my wet and dirty farm clothes and feel disheartened listening to the continuing rain and wind outside. I wonder "Will we ever relax again?"

When we started this farm, Casey and I learned quickly that everything is harder and more complex on our own. Even though we had experience, much of what we were doing seemed like learning from scratch — the weather included. While working for someone else, we learned the intensity of working outside. We brushed snow from kale on late-fall harvest days and felt the dirt and sweat caking on our bodies during hot summer weeding sessions. But, as farmworkers — not owners — weather was mostly just a matter of physical comfort.

On our own farm, however, the weather's vagaries determined our livelihood. Some days were seemingly perfect — mild, warm, sunny. As long as we irrigated, planted, and weeded, there was no question that we'd be harvesting crops. On other days, the weather changed our plans. Most of the time, unexpected weather was simply inconvenient: rain on spring days when we hoped to plant, for example. But weather could also swing to damaging extremes well.outside the range of averages and "normal." On our own, the stakes were infinitely higher. We had no other income — we needed our crops to grow and be marketable. We had to thoughtfully respond to the weather and make choices in order to keep the farm and our lives afloat.

SUMMER 2006

We've had so many heat waves this summer that I've lost count. At ten in the morning, I'm already sweating and realize that we're in for another one. After lunch, the temperature has risen to well over a hundred degrees. Casey and I start the irrigation on our fall broccoli planting and stand by, watching it wilt despite the water. We turn our attention to a weedy pole-bean planting. Steam rises from the ground as we hoe around the unhappy vines. Working in the heat is making my gut ache. Nothing we can do feels productive under the blazing sun, so we leave the farm and drive to a nearby swimming hole. We wade into the water in our work clothes and walk out again dripping and refreshed, even as our plants back in the fields are struggling.

Casey and I decided to farm in northwestern Oregon because it has a famously moderate climate. The promise of moderate winters hooked us into extending our vegetable-growing season and offering a year-round CSA program. However, since we moved here in 2006, Oregon's moderate weather has been punctuated by extreme events and record-setting on all counts — 2006 was the hottest summer on record, followed by the rainiest fall on record, followed in 2007 by the most powerful storm on record, and so on.

DECEMBER 2008

We're in the week before our final December CSA harvest when we hear rumors of a big winter storm. Knowing we can't harvest in sub-freezing temperatures, we pick our crops several days early, bunching greens during the few hours of thaw at midday. The day before our CSA delivery, the cooler full of bins ready to transport, snow starts falling. The next day we put chains on the tires of our box truck and drive the thirteen miles into town on icy roads.

We have met our commitment for the year, in spite of the challenges. Now we have to get through the rest of the storm and hope our overwintered crops survive for the earliest February and March harvests of the next calendar year.

That night, snow is falling again. And it falls and falls over the next week, totaling almost two feet, blanketing all the winter vegetables in the fields. Between rounds of snow, there's ice, which creates a hard sticky crust atop the snow, preventing it from sloughing off our greenhouses, which now bow and creak under the pressure.

Eventually the snow melts and the rivers fill. The waterway on our property rises inches daily. We watch as the water creeps into our neighbor's field and then over the bank of the creek on our land. Finally, we wake up and see that our lowest fields — beds of cabbage and other crops we hope to harvest in just a few weeks — are submerged under a clear river of flowing water.

With our anticipation of a moderate climate, the intensity of these weather events caught us unawares. Now, though, we have come to expect the unexpected. Here on our farm, extreme weather in every season is simply a fact of life. We'll never be able to predict what form it will take, but we're no longer surprised when the extreme arrives.

Until the world thaws again, we won't know that the three tons of potatoes we were storing in the ground will rot, that all but one of our cabbage varieties will melt from cold damage.

We've learned to prepare. We've learned that greenhouses must be set in concrete in order to survive strong winds. We've learned to plant diverse crops so that no matter in which direction the temperatures swing, something on our farm will thrive. We've learned that it still makes sense to turn on the irrigation and go swimming in the river on hot afternoons. And we've learned to relax. After weathering extreme events season after season and continuing to meet our commitments to our customers, we've learned that our farm is more resilient than we imagined.

DECEMBER 2009

We're entering another winter. We're bundled up in multiple sweaters, hats, and thick wool socks, walking slowly through our fields past frozen kale and mustard. The temperature has dropped quickly this December, reaching recent record lows. One night the thermometer reads 6 degrees, the next night 5.8 degrees, the next it's at 7. These are lower temperatures than many of our winter crops can survive, and we still have many harvests scheduled before spring arrives.

Until the world thaws again, we won't know what we'll have left. We don't know yet that the three tons of potatoes we were storing in the ground will rot. We don't yet know that all but one of our cabbage varieties will melt from cold damage. In fact, for the first time in three and a half years, I'm not thinking at all about how the weather may affect our crops. I'm busy working on our newest project — I'm pregnant with our first baby and in the midst of labor, walking laps around the farm to help bring on contractions. I labor through the cold snap, finally bringing our baby into the world on the third night.

A son! Casey and I rejoice at our good fortune as we count his toes and kiss his pink, glowing cheeks — new life born on one of the darkest and coldest nights of the year. When the temperature finally goes back above freezing during the day, Casey wraps our sweet baby against his chest and the three of us walk out to take stock of the cold damage. Yes, we've lost crops — more than we'd like. But spring will come. Spring will come. ✿

The Basket Is Half Full

BY TANYA TOLCHIN

With her husband, Scott Hertzberg, Tanya Tolchin runs a small diversified vegetable, flower, and herb CSA called Jug Bay Market Garden, about twenty miles outside Washington, D.C. She has worked on farms in Connecticut, England, and Israel, and also worked for ten years as an organizer and lobbyist for the Sierra Club. Tanya blogs at On The Lettuce Edge (see page 251).

It's a wonder that all farmers aren't multimillionaires. After all, we know the secrets of conjuring life from dormant seeds, of taking a single plant and transforming it into many, of turning dusty patches of weeds into beds of fresh marketable vegetables. Everyone needs to eat, a single plant can bare infinite offspring, and even forgotten perennials can bear fruit and flower. How can we possibly go wrong?

Sometimes farming feels that positive, and we need to remember these moments. Here are a few I remember from our farm. Last year, our non-irrigated potatoes grew so fast that it was actually absurd, almost cartoonlike. And an old bag of birdseed we threw on bare ground turned into hundreds of giant sunflowers. Our very first year, we had a completely accidental crop of very early beans. We planted them six weeks early, not to be clever but because, in our inexperience, we figured they were close enough to peas. The stars aligned with an unusually warm spring, and the beans were abundant and so early that we impressed the experienced farmers at our market.

Despite these occasional moments of blinding success, the books often just barely lean in our favor. Any farmer can tell you the list from the other side of the coin is long: the engines that fail you; the rain you are desperate for, that seems to pass over your farm and soak your neighbor's suburban yard; the deer that randomly develops a taste for tomato vines; the owl or the dog or the hawk in the chicken coop. When you win at farming, you get a small but hopefully fair wage for your time and work. But when you lose, you can lose big.

I have a friend who works in the dot.com world and has described the shop talk they use. Ideas go in baskets and need to be carried; they plant seeds of interest and harvest click-throughs and page views. It made me laugh because I know that many people find their work so intangible that they hide behind farm talk in their comfortable air-conditioned offices.

Years ago, my then-very-young niece asked me what I do at work. At the time, I was working for an environmental organization on forest protection, so I told her that I help save trees. She was excited, and asked if I plant them or water them. She was visibly disappointed when I had to confess that what I actually *do* at work was type on a computer, talk on the phone, and attend meetings. We could have any of a million different professions, and these are the same things we mostly *do* at work.

Most of us don't need to search for meaning in our lives; we see it every day.

Farmers work with plants, steel, soil, engines, and earth. We live by the weather, the earth, the seasons, the sun, and the rain. The baskets we carry have three dimensions. Most of us don't need to search for meaning in our lives; we see it every day. Thankfully, the work itself propels us forward to the next task. After seeding and planting toma-toes, you naturally want to tie up sagging vines and later pick the fruit. We know that money does not buy happiness, freedom, or meaning in life. And the farm brings all of these, some days in excess. ✄

Surviving Globalization Together

BY JANNA BERGER

A farmer and artist, Janna Berger has found many incredible homes around the world, including Angelic Organics, Red Fire Farm, and the Sunshine Farm. She owned and operated Living Stems Cut Flowers in Washington state and now manages vegetable and flower production at Adamah, in Connecticut.

I was a terribly angsty suburban teenager, but traveling abroad, exploring and transforming the foreign into the familiar, made the world feel real and worthwhile. In order to fund these trips, I donned khaki pants and endured work stints at places like Pizza Hut and the retail infant-supply superstore buybuy BABY. I was miserable doing customer service, but my paychecks were promissory notes for adventure. When I was abroad, I avoided the frenetic confusion and alienation of backpack tourism by volunteering, studying languages, teaching English, and working on farms. In this way, I happened upon the glamour of manual labor.

One such experience occurred during a reforestation project in Ghana the summer I turned twenty. I was clearing brush in preparation for tree planting and, like most of the other international volunteers, I was nowhere near as proficient with a machete as our Ghanaian hosts were. I quickly understood that our main contribution to the forestry effort was not our ineffectual hacking but rather the two-hundred-dollar program fee that paid for the saplings. Nevertheless, some combination of pride and sheer enjoyment brought me back to the brush each day along with the Ghanaian volunteers and a similarly persistent Frenchwoman, sweating and swinging my machete with little skill but lots of vigor. A stark contrast came into focus: the fluorescent buzz of my dreadfully boring days scanning teething-ring bar codes compared to the invigorating effort of clearing brush for replanting.

As such physical work experiences piled up, a new vision of myself as someone capable of using my body to alter the world emerged. With this realization also came a whole host of questions about what forces, if not my own self, had been shaping the world around me up until then. Suddenly, the effortlessness of the survival techniques I was brought up on scared me. Send in rent check: sheltered. Turn up dial on thermostat: warm. Pick up phone and order fried rice: fed. How did such a system work? And what if it didn't? And where did I, the fledgling machete wielder, the naive and enthusiastic novice, the green student of the international language of hard work, fit into all of that?

When I was twenty-two, I fell in love in Wisconsin. Thus motivated to deny my expatriate tendencies and find a job stateside, I took an internship on an organic CSA farm and quickly became somewhat of a farming nerd. I was thrilled by the daily work of bunching cilantro in the pouring rain, stuffing hundreds of infinitely squishable kale leaves into a single bin, digging up leguminous cover crops to see the mycorrhiza nodules at work fixing nitrogen on the roots, relieving myself in the brilliant technology of a compost toilet, and, perhaps most of all, seeing strong-armed women my size lifting and hauling and digging and creating with their bare hands.

The more I read and the more I pondered, the more I came to understand that local farms not only provide feverish excitement for the agronomically obsessed, but they are also a powerful offset to corporate hegemony over the global economy. Chicago salads are going to get their tomatoes somewhere, and it's better that they come from community-oriented organic farms in Wisconsin than worker-exploiting, chemical-soaked megafarms in Mexico and shipped thousands of miles on the same oil that fuels today's explosive wars and global climate change.

Across the United States, the number of American consumers buying local in order to decrease their dependence on the anonymous, exploitative, globalized economy has grown rapidly since I began my journey in small-scale agriculture six years ago. The broadening market for locally grown, organic goods has increased the viability of my making a living as a farmer, however modest by American standards.

This remarkable opportunity for me to begin farming is happening at the same time that doors are slamming shut on the opportunity to continue farming for most of the world's farmers. Over the past two hundred years, agricultural technology and the global food trade have rapidly pushed out small-scale farming and local markets. Industrial agrarian systems that use tractor power, hybrid and genetically modified seed, chemical fertilizers, pesticides, and crop specialization are highly productive, thus setting unprecedented low food prices. This industrial model requires expensive capital investments and land acquisition, making its adoption an impossibility for most farmers in the world and spurring massive rural exoduses into cities across the globe. Unemployment and poverty are often the result, as our economy simply hasn't created enough alternative professions to replace farming as a livelihood.

A new vision of myself as someone capable of using my body to alter the world emerged. What forces, if not my own self, had been shaping the world around me up until then?

Having totally restructured where and how humans live, the rock and the hard place between which industrial agriculture leaves us is even tighter, because its rapid consumption of resources makes it viable only as a short-term system. Topsoil is washing into the oceans due to erosion. Soils are so stripped of fertility that larger quantities of manufactured fertilizers and water are continually necessary. Herbicide-resistant weeds are emerging. Pest pressure on monoculture keeps ramping up.

We can't go backward to a time when the wisdom of one's ancestors could create a modern livelihood, because the environmental and market forces against it are too great. Surrounded by such a crisis and disconnected from the more sustainable systems that preceded it, where should agricultural pioneers look for modern agrarian models that reclaim agricultural livelihoods and soil fertility?

As a beginning farmer searching for the most sustainable systems, I have relied heavily on the knowledge and perspective of resurgent small farmers across the United States. Learning from farmers in other countries, especially ones who still have a relatively strong local economy, has proved more difficult. In a day and age when a Chilean avocado can make it onto a Canadian burrito before going soft, is giving voice to agrarian wisdom from the far reaches of human experience too much to ask?

Perhaps it is the nature of farming that makes worldwide networks of farmers educating one another and innovating together so elusive. With farmers necessarily engrossed in the universe of their own farm, it's difficult to connect. When the cucumbers need to be irrigated or the winter squash needs to be brought in before frost, it's easy to forget how big the world is.

My hope, however, is that in choosing to be a farmer, I have not relinquished my role in cross-cultural interaction, a resource too valuable to be squandered by individuals and dominated by transnational corporations and governments. In addition to minimizing the negative impact of exploitative global interactions, farmers, too, can reap the benefits of our modern capacity for global communication: We just have to open our eyes and seize the opportunities available to us.

My income and schedule as a farmer has made globe-trekking harder than it was when I took orders for Pepperoni Lovers pizza with extra cheese between flight departures, but I did manage to spend a winter in southern India a few years ago. There I met Indian farmers, their colorful, complicated farms similarly, stubbornly, tucked among mega-monocultures, whose environmentalism, humanism, and insistence on commonsense wisdom echoed those of so many American organic farmers. I was introduced to technologies I'd either never heard of or never seen in action, such as composted humanure, algal fertilizers, and the remediation of saline soils by certain types of trees.

Again, I was blown away by the way foreignness softened into friendship and the real potential for collaboration became clear.

Not all such lessons have to occur so far from home. Working on farms over the years, many of my coworkers were immigrants from countries where rural exodus happened more recently than it has in the United States. Often, their knowledge of their traditional farming systems goes unrecognized as they are relegated to laboring rather than creative management roles, an unfortunate situation that farmers can easily avert by having deeper conversations.

Networks of farmers have tangible things to share with one another, from heirloom seeds to effective crop-rotation schemes, not to mention the confidence to be gained from interacting with like-minded colleagues. Telecommunication, the Internet, books, air travel, and bold, basic listening are all potential tools for sharing strategies about how to survive as a small-scale farmer.

Most farmers cannot viably make a living solely by employing the methods of their grandparents, and yet the contemporary industrial model will destroy us all in short order. Founded in an epidemic of poverty and the environmental degradation that exacerbates it, the commercial bounty offered by globalization continues to float methodically over checkout scanners. The upside is that globalization also offers unprecedented opportunity for connection. In the face of the modern dichotomy that pits rich against poor and technology against tradition, today's agricultural pioneers cannot afford to leave the powerful tool of global communication unutilized. ✄

Lost and... Still Lost

BY BEN SWIMM

The manager of Spring Creek Farm in Palmer, Alaska, Ben Swimm grew up navigating the crowds at the downtown farmers' market in Madison, Wisconsin, where his favorite items were Amish sticky buns. Although he now attends markets as a vegetable farmer, he's still known to trade some of his fresh produce for the occasional danish.

Last season was my first comanaging a CSA, and it brought so many unexpected challenges that by October I was ready to collapse in a permanent heap of worn Carhartts and dusty long-sleeve shirts. With the season over, I kicked off my boots and banished farming from my thoughts and actions, wondering if I would return to it at all. It wasn't until I read an article by Verlyn Klinkenborg in the *New York Times* that I started to feel a bit better about farming. The essay, "Lost and Found," describes the author's struggle with the sometimes chaotic nature of farming. Especially frustrating to him is his bad habit of losing tools in the middle of a busy day. I have the same problem, so I connected with the essay instantly. To me, losing tools is a symptom of the larger challenges of farming, part of a chaotic spiral that's especially dangerous for beginning farmers.

It starts with the overambitious nature of most farmers, the inescapable tendency to pack more into one season than a single person could reasonably handle. This leads to flutter and haste, as one attempts to achieve more tasks than there is time for. This, in turn, causes disorganization and chaos. Combined with the resulting lost tools and time, it's easy for unanticipated events to spin this disorder into a dizzying cycle of accumulating and unsolvable crises that can eventually leave a farmer demoralized and helpless.

Further contributing to this feeling of helplessness for a first-year farmer is the seasonal nature of farming. Many problems that arise throughout the season simply cannot be solved that year, and damage control is only so uplifting. If your garlic doesn't sprout in the spring, you don't have another chance to plant more garlic that year. All you can do is buy garlic from the farmer down the road, wonder why you hadn't secured your mulch against the Alaskan-winter wind, and face the terrible reality that you had just one shot to grow garlic and you failed. After a number of these failures in the same year, coming back for another year doesn't seem appealing.

What makes it all okay, says Verlyn Klinkenborg, is that you usually find what you've lost, albeit only when you're no longer looking for it. Sometimes your lashed-together solutions, against all odds, do work. The satisfaction of these moments makes everything worth it, he says. The moment of stillness and clarity after his inspired solution for thawing out the water supply for his horses is triumph. His reaction to finding a long-lost object is idiotically joyful. These small victories are what keep him coming back for more. It sounds wonderful.

My response to Mr. Klinkenborg is this: That's easy for an experienced farmer to say. As a young farmer, though, you can't realize the foolishness of your quick solutions because you haven't failed yet. You won't find what you're looking for because you haven't been around long enough for it to turn up. Sometimes there's no positive resolution, no chance to get the better of a wind-ravaged greenhouse or a moose with a ravenous appetite for fresh peas and beets. At the end of the season, after emptying our workshop of all usable tools and leaving them in some field or box or drawer or greenhouse to be found some years from now, I had yet to feel a moment of redemption similar to Mr. Klinkenborg's. Sometimes pliers remain lost, or by the time they're found they're rusty and nonfunctional. Sometimes you can't find the right solution to anything. At least not on the first try.

What Mr. Klinkenborg has that I don't, I realize, is experience: years of tinkering that have given him an instinct of what will work and what will not, as well as an accumulated patience that has shown him that if he keeps working at it, things will eventually work out. When something goes wrong, he's been around long enough to know that he'll get another chance next year, and he'll be better prepared. One day I'll have the advantage of a similar perspective. In the frustrations of my first season, though, it was difficult to believe that farming would ever get any easier. At the end of an especially taxing season, this doubt made it hard to want to come back for another.

In October, I didn't know if I would make it over that hump and into a new year, excited and ready to farm again. The wisdom of an older farmer, though, tells me that the hurdles will get smaller each year. Next season will bring its own challenges, but I have to trust that it will also bring more of the transcendent moments that make farming worthwhile. This hope is what finally revved my excitement for the coming season.

```
What makes it all okay is that you
usually find what you've lost, albeit
only when you're no longer looking
for it.
```

Shortly after reading that essay in the *Times*, I picked up a book on composting. Soon after that, I began reading about soil science and discovered why our onions had thick necks and didn't cure correctly: Our soil is low in phosphorus. By mid-December, I was once again fully consumed by farming, reading, studying, trying to solve the problems of the previous year. What will bring me eagerly back to the fields in the spring is faith that the solutions are out there somewhere, and that someday I'll find them. Along with that pocketknife. ✍

CHAPTER EIGHT

OLD NEIGHBORS NEW COMMUNITY

My lettuce starts were sizzling in the field. It was May of my first farming season and I was trying frantically to get my new irrigation system up and running in the middle of an unseasonable heat wave. My desperately needed income was wilting in the field. The heat had come a month sooner than anticipated and I wasn't ready. I hadn't sourced my overhead sprinkler hand lines yet, much less finished burying the mainline, and half of the field was already planted — and thirsty.

I was sick as I swabbed PVC glue onto the last of the pipes, sweating behind the rubber respirator. The crop failure could be monumental. I might have to start over. I would lose half the season, half my projected income, not be able to pay off the startup debt I'd put on my 0 percent credit card before the no-interest grace period ended.

I looked up to see our neighbor Allen sauntering down the farm road. I knew him in the limited sense that his family owned the property next door; his cows were always busting through the fence onto my mom's pasture; he worked part time as a logger; he drove an old white F250; and his nephew, whom I'd always avoided on the school bus, liked to shoot peepers in the wet pasture swales each spring. Allen was in uniform that day: a shredded hickory shirt, a pair of suspenders holding up his ankle-length jeans, and some battered leather logging boots. He smelled of sweat, chain-saw exhaust, and Copenhagen.

"Looks like you need some water on that there field," he said, squinting into the afternoon blaze.

Did I ever.

"You got any need for some aluminum irrigation pipes?"

My heart jumped. "You have some to sell?"

"No. But I got some you can have."

The next morning Allen picked me up in his F250, towing a forty-foot pipe trailer. We spent the day pulling forty-foot irrigation pipes out of a blackberry thicket, from under a towering woodrat nest. He'd brought a one-inch pipe tap to rethread the corroded sprinkler outlets, and by mid-afternoon we had twenty workable pipes on the trailer.

Back at the farm, Allen helped me shoulder the pipe out onto the field. We hooked it up, turned the valve, and watched as a thousand leaks sprang at each pipe joint where the brittle gaskets were failing to seal. It was enough, though. Just enough pressure, enough water, enough mercy to save my ass that spring.

Fast-forward three years and the farm is thriving, with a hundred CSA members, a waiting list, and new gaskets in all the hand line. Allen comes to all our potlucks and farm softball games. I drop pints of strawberries and bags of potatoes — the only vegetable he likes — at his doorstep. I'm a new mother, juggling springtime madness that now includes a two-month-old baby and this book project. Meanwhile, we've outgrown our twenty-by-forty-foot propagation greenhouse, so I've decided to add on another twenty-odd feet. I'm pondering how to get the holes augered, the legs set in concrete, the ribs put up, the plastic stretched, the end walls built — all with a baby in the front pack.

Enter Tom, one of our founding CSA members, who throws himself into the project. He is retired and has been offering his help for two years, ever since he became a member of the farm. His refrain every week when I see him at the CSA pickup: "If you ever need anything, don't hesitate to ask." I have hesitated to ask, because I haven't known how to ask; have been too shy or proud to ask; maybe I don't want to ask for too much.

Suddenly, though, I'm in a place where I truly need help. Tom brings his tractor to the farm to auger the holes, and hauls every sixty-pound bag of concrete from the truck. Every day for five days he arrives with all of the tools we need, and together with my family we have a modern-day barn-raising of sorts. In less than a week, there's a new greenhouse, Tom has become a regular at lunchtime, and baby Cleo is cooing at him like he's Grandpa.

Old neighbors. New community. And one little farm in the middle of it all.

— Zoë Bradbury

The Farmers' Table

BY SAMANTHA LAMB

In yonder meadow, in the small farming town of Hobart, Oklahoma, lives farmer, writer, lover of picnics, and photographer Samantha Joelle Honey Lamb. She milks her cow, tends to her gardens, and creates art on a farm fondly called Early Bird Acres.

"Yes, that's how to permanently remove a tree stump, Lavern. Just put a tire around it, chip a deep hole in the center of it, put some gasoline in there, and burn the dang thing out."

This is what I hear coming from the long, sixteen-person table at the front of the diner in my new-to-me, but well-known by most, town of Hobart. It's a tiny nook in the wheat plains of Oklahoma. It's a happy and warm town, so the diner is appropriately named the Kozy Diner. It's the only breakfast place in a long way from the farm I live on and call Early Bird Acres. But this morning I'm not thinking about the raised beds I need to build or the twenty or so packages I need to bundle and ship. I'm thinking about how much I want to sit at the very lacquered and solid cedar farmers' table.

Nobody would look at me, a twenty-five-year-old girl who wears her hair in Dutch braids, and think "farmer." I do wear vintage gingham dresses and drive an old white Ford pickup around town. I do blast bluegrass music out its windows and try to convince random townsfolk of the importance of their baked goods being made with fresh ground wheatberries. But not a single one of those old farmers had ever looked at me as anyone near a farmer, and I don't blame them. Mostly I just come off as pastoral or bucolic. I've adopted the nickname "Swiss Miss," which is whispered as I pass by. Once I yodeled for their satisfaction.

When looking at careers, I always knew I wanted to be a farmer, so I decided to try to conquer the world first through art and then save up money for my own homestead. Once I finished college, I moved my dog, Harold, and myself to one of the family farms just outside of Hobart (which locally is pronounced Hobert). But the one person I needed most in this time of transition was no longer there. My grandfather, whom I lovingly named my dog after, was my inspiration for farming and bringing the local soil back into people's homes. The way he talked about the land was something of Sweet and Light. He taught me to see everything — from self-sufficiency to the small green sprouts emerging from the soil — in a way that captured my dreams and guided my actions.

The Kozy Diner eclipsed my homestead breakfast time by having the farmers table, and with it an invaluable amount of information.

The old men at that table had no idea that each one of them took on qualities of my grandpa as they spoke about good rain and tomato worms.

I didn't need to eat at the Kozy Diner. I had a whole mess of eggs, fresh spinach, and various other garden items right in my very house. I had a lifetime supply of coffee in my cottage kitchen cabinets, thanks to my friends back in Oklahoma City. But the Kozy Diner eclipsed my homestead breakfast time by having the farmers table, and with it an invaluable amount of information. I wanted to ask them "Where can I buy local seed? What's a natural way to treat the mites on my chickens? Who sells raw milk around here, or, better yet, who's selling good and healthy Ayrshire cows?" Alas, I didn't have the courage to simply walk in and sit at the table of elderly farmers in their overalls and well-worn suspenders.

One morning, after losing two of my chickens, possibly to coyotes, I finally mustered the courage to talk to the table. Not only was I distressed about finding just parts of the Rhode Island Reds, but I was also losing faith in myself as a farmer. What I didn't realize then was that communication with the outside, knowledgeable world was essential. Not just something I wanted, it was something I needed. Even my grandfather had to learn from his elders.

The previous week I had heard them discussing the coyote problem, and I thought this would be the perfect in. I wore my favorite cream-and-brown gingham dress, which, yes, twirled, and brought in some eggs for the waitresses. The giving of eggs was something I did out of nervousness. I had a habit of giving a dozen eggs to anybody I intended on meeting. The chickens were a perfect way to break the ice.

There was only one problem when I walked in: The table was full. With courage that came maybe from three cups of coffee or perhaps the warm smiles on the farmers' faces, I created my own place by adding a chair.

"Hello, folks, my name is Samantha Lamb. I'm new in town and I have a coyote problem."

"You're not that new, sweetheart. We've seen you coming in here the last three months and all you ever do is smile. Why did you never come over before?" a man I would come to know as Ernest said, as he slid me an empty ceramic mug. And then a man I would get to know as Parson poured me some Kozy Diner coffee.

The men thought I was a character of sorts, possibly from the realm of Disney. They routinely asked if the mice had sung to me that morn or if the birds helped me with my gardening. I informed them that the birds ate my strawberries. I was pleased and honored to find that they admired my fervor and dedication to my little farm, and they encouraged me with every tip and opinion. They might all have different opinions, but at least that gave me variety.

As a whole, the table was something of a wonder. Among the sixteen or so men at the table, they had experienced almost everything when it comes to farm life. I no longer felt foolish for not properly latching the pasture gate and letting all but one cow (the one that was asleep) out onto the wheat, because more than half the people at the table had done the same thing. When I had a good season of produce, we rejoiced. When I found the long-anticipated baby cow stillborn in the pasture, they were there to tell me I wasn't the only one, even though on some days I felt cursed. When I had to ask one of the men to come over to show me how to properly milk a goat, he did so without hesitation, even though there was a good deal of laughing. Also on that trip to my farm, he tried to demonstrate how to attach my donkey, Fred, to a plow, but all he got was an animal that preferred to nuzzle his pockets for apples rather than till the land.

Nowadays, I don't frequent the Kozy Diner as much as I used to, because life here at Early Bird Acres is very busy. I do still look to my farming friends for help, and ask them various questions about the garden and chickens, but now it's from the comfort of my own dining room table, made of walnut. This is what happened when they discovered that this female farmer knew how to make a good pie, and was willing to feed any hungry farmer in exchange for a bit of good Grandpa advice. ✍

Buried Steel

BY SARAHLEE LAWRENCE

In Terrebonne, Oregon, Sarahlee Lawrence runs Rainshadow Organics. She has also started a farm school, called Plots to Plates, and is the author of *River House*.

I grew up on a hay farm in central Oregon. As an only child, I spent a lot of time with my horse and hauling hay for my dad. When I moved away to college, I had no intention of coming back. I became a river guide and let the river carry me around the globe. Somewhere along the line I realized I had become untethered. Like a balloon floating up out of a little girl's hand, I looked back at the family farm with new perspective. With the realization that food is everything, Monsanto was everywhere I traveled, and my dad was ready to turn over the farm, I headed home.

I don't know what made me think I could start an organic-vegetable farm and CSA. I didn't even eat vegetables. It had been ten years since I'd lived on the farm, our climate is notoriously difficult, and I had never planted a seed. Never mind that. I went wild with the seed catalogs, put a deer fence around two acres, buried a drip irrigation system, and built a great big greenhouse. At the end of March, it was time to "raise" the beds.

The ground had dried up just enough and a friend and I set to the field with shovels in hand. After a couple of grueling days and twenty-nine thousand of the thirty thousand row feet still left to raise, it was clear we needed to work smarter.

I live in a valley of farms — mostly hay and grain — that are on a much larger scale than my potential veggie patch. I've known my neighbors forever. Not much changes out here. The old farmers get older and we still get together every year for Christmas. Even though it had been a decade since I'd seen most of them, they took me back into the gritty fold as if I'd never left.

In the case of the bed raising, I knew just who to talk to: Glenn Cooper. He'd been farming in the Lower Bridge Valley for half a century. Glenn lived four miles away, at the end of the road. I found him in his shop, full of old engines that he spent his spare time rebuilding. Once he'd built an entire car from random old parts. He called it the Cooper. I poked my head in the door.

"Glenn?" I called out in my sweetest voice.

He looked up from his work and stared at me through his big square glasses, pocked from running his welder. Glenn lived about as far out as you could get and was well known for shooting at trespassers on the gravel road beyond his gate. That is, if his dogs didn't flatten their tires when they stopped briefly to consider the NO TRESPASSING sign. I was always a little scared of him, but he cracked down on the cheap cowboys in black Bailey hats who pilfered hay from our barn in the winter, and we appreciated that.

"Katie said you needed a tool," he said without any particular kindness.

I had called ahead to warn him I was coming and had talked to his lady friend on the phone. "Sure do," I said as I headed in. "Something to raise my beds."

"For those vegetables?" he asked. "They're not likely to grow here, you know."

"I know," I said, brushing that off. "I'm breaking myself trying to raise my rows."

"I know," he replied, "I've seen you working at it." Glenn put on his old greasy green trucker's hat and his jean jacket. "I've got just what you need." He stepped out of his shop into bright sun and opened the passenger door to my pickup. "It's out in my junkyard."

Glenn's "junkyard" was several acres of equipment parts and hunks of metal. We trolled around in low gear between the juniper trees and sagebrush, glancing out the windows at the riches.

"Now, when I moved here forty-five years ago from northern California," he said, "I had a couple of kids, five hundred head of cattle, and about five hundred bucks. I brought up two opposite plows that we could mount to a tool bar and hook to the three-point on one of my old tractors." He rolled down the window and stuck his head out. "I'm pretty sure I set them together out here under a tree."

I raised an eyebrow and looked at Glenn, who was looking out the window. I didn't really know what we were looking for and I couldn't believe that anything that had been sitting for forty-five years could be of much use.

"I've got just what you need. It's out in my junkyard."

"Oh hey," he hooted. "I think I see them. Just stop right there."

He dropped down out of the pickup and strode over to a big old juniper tree. I left the truck running and caught up.

"Yup, here they are," he said as he kicked at the steel. "They're buried."

"Well, what do you think?" I kicked at them too, and they didn't budge.

"Back your truck over here and we'll pull them out with the chain in the back."

Luckily I had a chain, which saved me much embarrassment. I still couldn't believe we were going to do anything with these things, whatever they were, but I backed up the truck anyway. He hooked them up to the bumper and I eased forward, pulling the slack out of the chain. When he signaled to stop, I stepped out to take a look at what we'd unearthed.

"They're huge!" I exclaimed. Three feet long, scooped steel, clearly for moving dirt.

Glenn didn't react to my surprise. "Well, help me get them into the truck," he said.

We got our hands on the awkward things and hoisted them into the bed. We rolled back to his shop and he disappeared into the bowels of the building, only to emerge with a giant steel bar rigged to be hitched to the back end of a tractor. He went back in and then reappeared with a giant wrench and some solvent to clean up the plows and free the old bolts.

We spent about an hour getting the plows on the tool bar and hooked to one of his many old tractors. Unexpectedly, he clambered up into the cab of the idling tractor, trundled over to a patch of bare ground, then came to a stop. He eased the plows into the dirt and I watched as Glenn's rusty steel contraption sliced through the ground and folded it effortlessly together into perfect rows almost a foot tall.

"Yee-haw!" I squealed, jumping up and down.

He eased to a stop, dropped the RPMs on the tractor, and stepped down. I ran to him and threw my arms around his shoulders.

"I love you, Glenn!"

"I know," he said stiffly, standing there trapped in my bear hug.

"You can take the tractor," he said tersely, then looked me in the eye. "But leave your truck and bring it back as soon as you're done." ✑

It Takes a Village to Raise a Farm

BY JON PIANA

A farmer, fermenter, and community activist in Barnard, Vermont, Jon Piana operates Fable Farm with his brother, Christopher. His guiding mission in life is to live within his means.

It was October, and we were buried in potatoes. It had all started the winter before. Our root cellar had already been stocked deep with seed potatoes from the prior season's harvest, but we had been tempted by the mouthwatering descriptions that the authors of the Fedco seed catalog had cleverly written for each of their varieties.

In midwinter we impulsively ordered more, which meant that by the time the spring dandelions showed their yellow crowns — telling us to get our taters in the ground — we were swimming in seed potatoes. Maybe we'd been guided by an innate survival instinct: Potatoes are meals in themselves, and in a land where winter settles in long and heavy enough to leave most life processes dormant, a potato crop is crucial to sustaining a local foodshed. Or maybe it was simply the promise of pleasure: Up against the cold, dark days of a Vermont winter, potato flesh moist with melted butter is a delectable comfort.

No matter what, come October we had more potatoes in the ground than my brother, our two interns, and I could manage to harvest with just our hands before the snows came and the ground froze.

A walk-behind BCS rototiller is the largest piece of machinery we use to tend the four-acre patchwork of borrowed farmland peppered throughout our hillside village. It was this machine that was responsible for digging the furrows into which we dropped countless spuds. This year we had too many rows to hill by hand with a hoe, so we bartered with old Neil Campbell for the use of his tractor-drawn potato hoe. When it finally came time to harvest the fifteen hundred row feet of storage potatoes, our saving grace was not the mechanical power of the tractor, but rather the many hands of our fellow villagers.

In the spirit of old-fashioned harvest celebrations, we called a Potato ShinDIG: "Come help us reap the harvest and take home your winter supply of taters. Revel in the soil with friends and cider on tap. May we get down and dirty."

I arrived late to the ShinDIG, because of a last-second snafu with the keg of cider at our farmhouse. By the time I reached the potato field, I was expecting the digging crew to have just begun. Lo and behold, most of the spuds had already been unearthed, and folks were chatting casually as they dug. It was warm for a late-October afternoon in the hills of Vermont, but the soil was cold and wet, leaving the digging crew swathed in mud. I stopped for a moment to take in the scene: bodies of all shapes, sizes, ages, and abilities bent in varied poses, fishing food out of the ground.

Before all the potatoes had been gathered into baskets, the cider was tapped and a celebratory spirit permeated the group. We stood around, covered head to toe in black earth, sharing the cider from last year's apples. The setting sun cast its glow over our village as we reflected on what a rich and fruitful year it had been for the town of Barnard. What would have taken us weeks to harvest, our community had done in an hour.

My brother and I were born not on the farm, but instead at the intersection of rural and suburban culture. Our family made annual trips to local farms for our pumpkins, Christmas trees, corn on the cob, and apples. Beyond these quintessential farm visits, though, it was rare to have a hand in agriculture. Our upbringing was dominated by academics and athletics, both of which demanded discipline and a work ethic that unwittingly prepared us for our unlikely future as farmers.

I graduated from college with a liberal arts degree, my brother turned down a grad school program, and our collective commitment fell on the soil in blind faith. I actually didn't know where the village of Barnard was before November 2007, when Joe (a beef farmer we lease land from) carved out a two-acre vegetable plot for us on his farm. This newly upturned loam represented the boldest decision we had ever made. It wasn't long before Farmer Joe became Grandpa Joe.

That February, we built our greenhouse, digging through four feet of snow to reach the ground. Meanwhile, we spread the word about our vegetable CSA venture across billboards and in local papers and sought a place to rent. By the thaw of '08 we had found a place to live in the village center and developed a twenty-five-member CSA called Fable Farm. We moved to Barnard knowing not a soul; what we discovered was a community eager to help and a village boasting a rich agricultural past. Thanks to our scavenging abilities and thrift, and thanks to the CSA members we'd recruited that winter, we didn't have to approach a bank and thus avoided sinking into the black hole of debt.

Now, three years later, we offer a hundred vegetable shares and sell to select markets. We don't own any land, but we do have increasing access to rent-free farmland throughout our village, thanks to the generosity of some townspeople who recognize the severe shortage of available land and want to see our enterprise survive. After three growing seasons (with winters off), my twenty-six-year-old self has been able to save some money. Confronted by the classic hurdle of inflated land prices, it's not cash that will buy us land to tend in perpetuity, but rather the currency of human connection. I no longer question where I should settle, for it would be sheer stupidity to walk away from the networks of friends and supporters here in Barnard.

In the beginning, though, we needed to prove to the townspeople that we were capable of growing good food and that we were not afraid of hard work. We were blessed that first year to be farming on Grandpa Joe's well-cared-for, rotationally grazed beef farm. With the grace of the fertile loam below our bare feet, our vegetables grew in abundance. The only complaint we got from our CSA members was that we gave them too many veggies.

The second year brought us more into the town. Cherishing his privacy, Grandpa Joe told us we needed to find a new site for our CSA pickup days. It just so happened that an old farmhouse sitting on half an acre in the village center was for rent. Its owners were looking for tenants willing to work on the house in exchange for reduced rent. They permitted us to plow up the backyard and use the space for our CSA pickup. Despite the fact that the length of our lease was uncertain, we built an earth oven in the backyard — a clay temple that's becoming the hearth of our village.

What would have taken us weeks to harvest, our community had done in an hour.

Barnard is a rural town that reached its cultural peak in the late 1800s, then declined with the broadening of a globalized economy. The population of nineteenth-century Barnard was double what it is today, and everyone had a hand in the production of goods, whether by farm or factory. The lack of mechanization required people to work together in order to reap the harvests, thresh the grain, and bale the hay. Their survival depended on community. As small farms and cottage industries failed or left, what it meant to live in community began to shift. The local gathering places closed their doors as fewer people were making their livelihood in town.

Still, today this is a good place to raise a family, but when children grow up and need to earn a living, most leave for the cities. That is, until recently: With the recession, many are returning home to Barnard, to reinvent themselves and stimulate a rural economy.

My brother and I are some of those younger people who are migrating to our nation's small towns. We started farming not only to feed ourselves and to live within our means, but also because we saw it as a means of community service. Growing healthy food and building soil are important to us.

And so our CSA pickup is not just an exchange of vegetables. It's a place for children to run free through the nooks and crannies of the U-pick garden. It's a place where people can relax with neighbors over a cup of cider and some flatbread from our earth oven. It's a place to get together and celebrate the harvest, one another, and the changing of the seasons, all to the sounds of music coming from under an apple tree beside the hearth. (We host live music and other performance arts at every pickup, providing a venue for local musicians and artists.)

During our weekly pickups, the backyard becomes a free marketplace where people sell other agricultural goods such as raw milk and butter, fresh pasta, fermented foods, breads, and meats. On the pickup days when we don't bake flatbreads, we host community-wide potlucks, which go into the night and end around the fire.

You don't have to belong to our CSA to enter through the white picket fence into this garden in the village center. Here among the vegetables, flowers, herbs, and earth oven everyone is welcome, and it's not uncommon to find ninth-generation Vermonters mingling with second-home owners and farmers shooting the breeze with lawyers.

It's through these celebrations of food and agriculture that many people, some of whom haven't been in a garden since their childhood, develop a desire to labor on the farm. What starts around the distribution of vegetables wends its way into the fields where corn needs shocking or barley needs threshing.

So it was on that muddy October afternoon. There were vegetables to be picked up at the farmhouse but also heaps of potatoes to be dug and much to be thankful for. People were eager to sink their hands into the heavy soil, to be in touch with where their winter potatoes came from, and to pause in appreciation of community and the fields of plenty.

Social Farming

BY JEN AND JEFF MILLER

The owners of Dea Dia Organics, in Grayslake, Illinois, Jen and Jeff Miller recently joined forces with a mentor, Sandhill Organics. This is the latest chapter in their collaborative farming experience. The combined farm will comprise approximately twenty acres and produce organic vegetables and pastured eggs for families in the greater Chicago area.

We were drawn to farming because we love to work hard and eat well. Although our education is in landscape architecture, art, and marketing, we found our passion in farming. As with many other farmers, we didn't become farmers because we were looking to socialize; we were looking for solitude. We wanted to own a business that enabled us to slow down and enjoy our lives each day. We wanted to do everything from developing business plans to working in the soil. We wanted to figure it all out ourselves and "own" the process.

So, during the winter of 2005, we got to work networking, to discover how to get started. We learned the business side of farming through Stateline Farm Beginnings, a farmer-training program. We visited farms throughout southern Wisconsin and northern Illinois to learn from working farms. During this time, we sought a place to get our hands dirty, and were introduced to the idea of a farming incubator (a training program that offers access to land, equipment, and mentorship while you develop your farm business).

Our farm began its operations the following spring as part of the Farm Business Development Center (FBDC) at Prairie Crossing Farm. The FBDC, located in Grayslake, Illinois (forty miles north of Chicago), was created as a way to help farms get started in a supportive environment, without the large capital investment normally required. We share space, some equipment, and, most important, ideas. After joining the FBDC, we rapidly began to see the beauty of farming "collaboratively." We work alongside other farms, some with more experience, some with less. At first we were mentored by more experienced farmers; gradually, we became mentors to the newest ones.

Thus we began farming not in isolation but as a small group, members contributing a different set of skills, varied knowledge, and diverse resources. These farmers became our friends. Our collaboration dramatically improved our learning curve, and we embraced this social type of farming. Working alongside others also made us realize that we wanted to be more active participants in the community.

Community has always been an important part of farming. From barn raisings to bringing in the harvest, working together makes our job what it is — though challenging, it is also feasible and enjoyable. We like hosting social events at the farm: These events provide others with the opportunity to get to know each other, at the same time helping us transplant tomato seedlings, for example. Later in the season, we gather at the farm to see how those transplants have grown, then reap the harvest together.

Though nothing can replace this face-to-face time with our CSA and community members, some of our most valued relationships developed through an online dialogue. This made us wonder: How do we as farmers connect the people online to our delicious tomatoes? We've grown into farmers during a huge evolution in online media. We can engage with our fans, friends, and family while sitting at the computer. We provide a realistic picture of what goes into growing food, which generally opens up a dialogue about farming (and how it seems like a lot of work).

We're not just broadcasting; we're engaging and building deeper connections with those who support our farm. Each social tool has its purpose. For example, we use Twitter for quick shout-outs: On a Monday, we might send out a short note to let people know about a project we're working. Then, when members come for produce on Saturday, they ask us how the project turned out.

We also use Facebook to engage with our farming community. Once we posted a picture to see if anyone could guess what our newest homemade contraption was. The caption following the picture was "A dozen eggs to the first person who can guess what this is. Parts include a cordless drill, wheelbarrow, car battery, plastic drum, and a hose coming out the bottom." The responses ranged from a new kind of port-a-potty, a solar water pasteurizer, and a salad spinner to "Are you making butter?"

The conversation lasted for a few days, and involved eight people and many thumbs up. (The actual contraption was a side dresser for fertilizing crops.)

Our blog serves as our main tool for communicating longer stories, as well as our CSA newsletter. It gives members valuable information, and they're more supportive because they're learning something new. Like other farmers who have experienced major flooding or crop loss, we share how difficult weather affects our ability to grow the fingerling potatoes we hoped to offer that week. Our customers get a good sense of all the planning and work that go into each product, enabling them to more fully appreciate their food and embrace the CSA experience. Our best customers seem to be those who read our weekly blog, contribute to the conversation, and are eager to learn more about how we manage to grow such a wide variety of crops.

Though nothing can replace this face-to-face time with our CSA and community members, some of our most valued relationships developed through an online dialogue.

As farmers and farm business owners, we have a responsibility to both our online and our offline communities. We may work alone on a project during the day, but we always carry along the tools to capture the experience through pictures, videos, and posts. This is one way we share why we do what we do and engage people in a discussion about how and where their food is produced. This ongoing conversation not only benefits our online social community, but it also feeds our passion to be intimately involved.

On the farm we can escape the type of one-upmanship we used to feel (and dislike) in the careers we once had. We've realized, though, that we didn't have to achieve this through solitude. As it turns out, we weren't really seeking solitude as much as we were looking for a community that embraces hard work, good food, and sharing. ⌀

Who Says You Can't Go Home?

BY BRANDON PUGH

Brandon Pugh lives and farms on his family's land in Proctor, Arkansas. His farm is called Delta Sol Farm and includes livestock, vegetables, and flowers. He prefers being a part of renegade markets and eating good food, and he loves living in the Mississippi Delta, where he can hear boats on the river while he's farming.

I grew up and now farm in Proctor, Arkansas. My farm is called Delta Sol Farm, and it's located five miles from the Mississippi River and a twenty-five-minute drive from Memphis. This is river-bottom soil, so the land is flat and rich from years upon years of flooding. The summers are long, hot, and muggy, which makes this place perfect for quick-growing weeds, bugs, fungi, mildew, and more bugs. But you can also raise lots of crops. We have an amazing growing season; winters can dip down into freezing, but are usually mild. The heat of summer is what can drop-kick you in the face.

The majority of farmers in my area, including my big brother, grow soybeans, cotton, grain sorghum, rice, and winter wheat. It's big-time conventional ag with all the GMO-pesticide-subsidy-crop-duster junk we all just love. My family has been in this area for more than four generations, and I'm currently farming on an old four-acre pasture behind the house I grew up in. What I do is small potatoes compared to my brother and other farmers who have more than thirty-five hundred acres in production. It's pretty hard to imagine being responsible for all that land when my four acres keep me busy day and night. However, it sure is nice having a brother who farms because I can always count on him or one of his workers to help me out when something on my farm breaks.

I was able to go away for college, and received a degree in environmental studies with a focus on sustainable agriculture. From there I went to work and apprenticed on small, diverse farms from New

Hampshire to California and met some amazing farmers. My farmland back home was always in the back of my mind, though, and after spending five years in the San Francisco Bay Area, I felt it was finally time for me to go home. The resources and opportunities just couldn't be passed up. There was land to farm, housing available, and my dad would even let me use his small tractor. All these things are extremely important for getting a small farming business off the ground and making it successful. Also, this area of the country is very affordable to live in and the "buy local" trend is really catching on.

Being here means I get to create a sustainable farm in the middle of big industrial ag, and I can be a pioneer in the local food scene.

It was great to be back home around my old neighbors. There was support for my endeavor from the community I grew up in; it was sort of the talk of the town. Folks were excited that Brandon Pugh was moving home from "out West" to do "that organic-vegetable thing."

The first year I did my CSA, it mostly consisted of my mom's friends, who really just wanted to be supportive. I don't think too many of them were ready for all the produce coming their way. The CSA pickup was a time for conversation. Folks discussed how the bingo went last night, who was sick, what they were gonna do about that new road being built, and other important town news. Also, people enjoyed "shoulding" on me. And they still do. Everyone wants to tell you what you should grow, where you should sell and other brilliant ideas that you "should" do. They say things like "You really should grow two acres of okra" and "Have you ever thought about growing catmint?" I've learned to act like I'm listening and nod my head.

My family's resources were helpful in starting my farm, but it was still pretty crazy building a greenhouse, irrigation system, washing/packing shed, and cold room all at the same time that I was growing and marketing my first season of produce. It turned out to be a great time to know folks in my area who could come help me build structures and set up pipes.

There were still plenty of other challenges to adjust to, though. Farming without chemicals in the South can be a losing battle, between the weeds and bugs that'll take over in no time if you let them. In my first season, you could find me kicking my way through the Johnson grass to find my watermelons. Really annoying. The heat of the summer is something that can stop you in your tracks. Many times during the really hot days, I feel like I get some type of runner's high from the heat and my body just kinda starts to tingle and I really feel like I've smoked a pound of pot. You don't realize how hot you are until you take a break for a second. You feel like you started the day weighing one-eighty but then sweat away ten pounds. Also, there aren't many cool swimming holes around here; they're all shallow, snake-infested swamps.

The other challenge is that pesticide drift is a way of life. Seems like we all just grew up around it and never really thought too much about it. But it became a huge deal for me — and a very real moment in starting my farm — when I was sprayed by a crop duster. On both sides of my farm, there are large soybean fields. I know the guy who farms that land and when I returned home I had a conversation with him about how I was farming without the use of chemicals and that I'd like him to take all precautions necessary to avoid contaminating my fields. That seemed to be working fine, but apparently he forgot to mention it to the crop-duster company.

I was out weeding my carrots one morning when I noticed the sound of the crop-duster getting louder and louder. You can hear them coming and within twenty seconds they're on you. They fly all around here but this time he was getting way too close. I froze and didn't know what to do. Then the plane started spraying the field to the north of me. He wasn't right on top of me yet but I was pretty sure he was on his way. And sure enough, the bombardment started. I actually tried to keep my cool and keep weeding, but when I felt the chemicals touch my body, I freaked out. It was pretty horrible, such a helpless feeling. I had been working so hard on my farm and they came and contaminated it in less than five minutes.

I had a camera in my truck so I ran and grabbed it and started taking pictures of the plane. I think I also threw my hoe or rake up in the air at the guy. I got up on my truck, flailing my arms and giving him the "what are you doing!" look. And before I knew it, he was gone and all my hard work felt tainted. It seemed like it was all for naught.

I was outraged and ready to get some vengeance. The first thing I did was call the farmer and freak out on him. I was ready to protest and call the paper and sue everyone involved and pretty much just shame these folks in front of the community.

But of course that's really not the best way to solve problems. The farmer gave a genuine apology. He also said that it was only a fungicide and suggested that maybe it could help. And although I was still very angry, I realized that he truly believed it could help in some way. As much as I wanted to yell at him, I had to stand back and hear him. I'm starting to realize that many folks around here think that if it's not an herbicide, then there shouldn't be a problem.

Same thing went for the crop-duster company. I called and sure enough, having grown up in this small town, I knew the manager. I think we went to kindergarten and were in Scouts together. After I explained my situation to him, he too apologized and said he didn't know that I was there and that it wouldn't happen again. I want to believe him but I still freeze with panic every time I hear that crop duster.

As far as I can tell, the crop dusting didn't affect my health or any of my crops. To hold them accountable, I got in touch with the FAA (Federal Aviation Authority) in Little Rock. Now I have someone to call if they ever do it again. The farmer lets me know whenever they might be spraying. He makes sure the wind is blowing away from my farm. And the crop-duster manager has agreed to come stand in my field with me the next time they spray near me.

A huge lesson from the whole ordeal was learning how to communicate with folks here at home. It seems like you can get a lot accomplished by educating folks and informing them about why we farm the way we do. And if we aren't all on the same page, as farmers we still need to respect each other. It's easy to get caught up in thinking folks are "bad," but really, everyone is just trying to make a living. Farming is a hard business no matter how you do it. My brother is probably the hardest worker I know. For at least eight months a year he's out there working his thirty-five hundred acres from sunup to sundown. He has mouths to feed and bills to pay and this is how he can do it. And as long as they aren't affecting my business, small as it may be, then that's a first step in becoming a part of this community of farmers.

Odd as it sounds, it feels good to be here farming the way I am, because I'm not surrounded by other small organic farms. I'm out here alone, but I have my family's support and my community's support and they're seeing by my example that there are other ways to farm. I'm in the process of becoming certified organic, mainly so I can better enforce a no-spray policy around my farm. But also it's a great feeling to be here getting a new way of farming started.

I'm now in my third season of farming here, grateful and proud for what I have. I was able to follow a natural path that led right back to where it all began. I'm living in the house that I grew up in and haven't had a second thought about my decision to move back. Being here means I get to create a sustainable farm in the middle of big industrial ag, and I can be a pioneer in the local food scene. I sell at several markets and have also gotten to know some great chefs who are excited to have my produce on their menu. The CSA is growing among local families and friends, and the pickup (which is at my parents' house and staffed by my mom) is when we all catch up on what everyone's kids are up to or share a pickling recipe or a squash casserole.

My neighbor, Mr. Reynolds, and I see each other almost every day. I've known him all my life and grew up with his kids; now it's nice to be back outside around him. He retired the same year I moved back, so he's always out mowing his lawn or working on one of his tractors. We meet at the fence and talk about what crops I'm growing, what our other neighbors are up to, and what to do about these pesky crop-dusters. ✄

Cross-Pollination

BY LIZ GRAZNAK

A native of Columbia, Missouri, Liz Graznak's love of gardening is the result of time spent with her grandparents as a child. She discovered CSA farms in grad school, and, after interning and working with other farmers, she moved back to Missouri. She's living her dream of growing beautiful, healthy food for a community of people who share her enthusiasm and commitment to a local food system.

The summer I started my CSA farm, it rained twenty inches over the yearly average. Experienced growers jokingly said I was lucky to have challenging conditions in my first year because subsequent years would be easier by comparison. Though my bottomland at Happy Hollow Farm was saturated, and some of the organically grown vegetables I'd planted early in the season drowned, the tomatoes thrived. Most other gardeners in this mid-Missouri rural area lost their tomato crops to rot and disease, so my success didn't go unnoticed in the local community, particularly by my neighbor J.T. Cassil, whose family (which has been in the area since 1870) used to own my farm.

For most of his adult life, J.T. was a dairy farmer. At seventy, he still raises cattle and bales hay on the farm adjacent to mine. He drives a Jamestown school bus and has safely transported three generations of students to and from school each day. Now that he's retired, J.T. enjoys driving the students across the state for ball games, concerts, and field trips and sharing those experiences with them. He delivers gravel to folks in the neighborhood who need their roads improved or building foundations set. He's an enthusiastic banjo player in a gospel band; he's naturally talented and plays mostly by ear.

My partner, Katie, and I knew little about the Jamestown community three years ago when we moved to Happy Hollow Farm from Columbia, a university town of a hundred thousand just a forty-five-minute drive away. We discovered that our new community consisted of many local families, like the Cassils, who've lived and farmed here for more than a hundred years.

As soon as we got settled, I made a concerted effort to get to know my new neighbors. One of the first things we did was visit J.T. and his wife, Mary, with what turned out to be a terrible homemade cherry pie. J.T. and Mary were very gracious about the chewy crust. Mary suggested that lard would improve the flakiness and told us that their lard comes from hogs they butcher. That evening, not only did I learn that the best piecrusts are made with lard, but I was also reminded that the best friendships are forged by sharing experiences and spending time together.

Even though our mutual admiration grew rapidly, J.T. was often perplexed by my new-to-him farming practices.

I felt an immediate affinity with J.T. because he reminds me of my grandfather. His tough-love, hardworking attitude is one I was raised with and have come to embrace. When Katie is traveling, J.T. checks in on me and Mary brings over food, knowing that I have little time to cook. Last spring J.T., his cousin, and his two sons helped me cut down three huge locust trees along the side of our shared road and hauled them to the building site of my new barn. He introduced Katie and me to his friends at a local Labor Day picnic and at a church dinner, opening doors for us into our new community. He helped me change the tire on my old manure spreader, a life-threatening job if not done properly, and two months later gave me an advertisement for a newer, less dangerous machine. A week seldom goes by that I don't see J.T.; usually he's stopping in to see if I need anything, but sometimes it's just to say hello.

Even though our mutual admiration grew rapidly, J.T. was often perplexed by my new-to-him farming practices. Why was he helping me unroll large round hay bales over sod for what I was calling "permanent beds" where the first tomato crop would be planted? Why did I wait to put out my tomato plants until late May when everyone else sets out theirs earlier, aiming for tomatoes by the Fourth of July?

Farmers in this area traditionally raise corn, soybeans, and cattle while holding off-farm jobs to supplement their income. Most also use conventional growing methods, spraying synthetic fertilizers, insecticides, fungicides, and herbicides throughout the year. Organic farming is considered unconventional, "hippie-esque," and not economically feasible. Most local residents consider the type of farming I'm doing to be truck farming, the old term for selling produce out of the back of a pickup, and are unfamiliar with the concept of Community Supported Agriculture (CSA), in which the public pays the farmer in advance of the growing season for a share of the farm's weekly harvest, and shares some of the risk that farming entails.

In the spring, J.T. and Mary helped me put in almost three hundred tomato plants, taking some of the greenhouse-grown young plants for their own garden. As the tomatoes grew, we were all impressed with the lush green foliage and their ever-increasing size. As summer wore on and the rain continued, my tomato plants thrived while most other gardeners were losing theirs to blight and having poor fruit development on the vine. Once harvest began, there were so many tomatoes that J.T. again brought his cousin and son over to help pick, and Mary canned thirty quarts for Katie and me.

The twenty members of my first-year CSA received six varieties of tomatoes, and a few bought extras to can. My tomato harvest was abundant, but I had plenty of problems with other vegetables because of the cold spring and wet conditions over the summer. The potato and onion plantings were failures. I lost two of the early broccoli plantings, and most of the summer cucurbits and crucifers suffered from too much water and not enough heat for good fruit set. Though I was initially disappointed that I didn't get the fifty CSA memberships I'd hoped for, it turned out well enough, considering the adverse growing conditions, my inexperience with the new bottomland, and first-year projects we undertook: a greenhouse, a modified timber-frame barn, a walk-in cooler, and polytunnel funded in part by the new organic farmer cost-share program through the Natural Resources Conservation Service (NRCS).

In my second year on the farm, I grew a wider range of vegetables in hopes of keeping the weekly CSA boxes diverse and interesting throughout the twenty-five-week growing season. With all of the tomatoes, peppers, lettuces, cucumbers, and squash I grew, there were more than a hundred varieties of vegetables that found their way into the weekly shares. Some were new to my members, such as escarole, tatsoi, fennel, 'Hakurei' turnips, celeriac, and what turned out to be J.T.'s favorite discovery, kohlrabi.

J.T. and Mary were surprised by the variety in the produce I grew. They seemed impressed by my ability to do most of the farming by myself, and by the end of the summer they attributed my success, especially with the tomatoes, to my organic practices of using mulch and compost. In October, for the first time, J.T. unrolled hay onto his garden in preparation for next year's season. Over the winter, J.T. and Mary's soil will be protected and the earthworms will rise to the surface, loosening and adding organic matter to it. As the mulch decomposes, nutrients will be added back into the soil, and if all goes well, J.T. and Mary will have a bumper crop of tomatoes.

The cross-pollination taking place between J.T. and me fulfills one of my major goals in life: to meet and befriend new and different people. It's one of the reasons I decided to start a CSA. I wanted to grow healthy, chemical-free food for people, and to forge wonderful friendships and a support network to aid my farming journey. Even though J.T. and I do encounter certain limits to our relationship — J.T. didn't attend Katie's and my commitment ceremony and despite his invitation I choose not to attend his church — our friendship remains strong because it is based on respect for each other, for hard work, for caring for the land, and for nurturing family and friends.

Over the past two years, our rural farm community of old and now new farmers has grown and thrived. As J.T. often says to his neighbors, "Any two women who work as hard as they do are okay in my book." ✍

Coming Full Circle: The Conservatism of the Agrarian Left

BY VINCE BOOTH

After farming in Nevada City, Vince Booth returned to the familiar clime of eastern Washington. He grows and pickles for Booth Brine Co., a live-cultured pickling operation, in Walla Walla.

In fall 2009, two fellow farmers and I attended our district's town hall meeting about healthcare reform. As you may recall, the national debate at that time was so feverish that people were brandishing rifles and biting one another. The meeting we attended was theatrical when not a battlefield. By the end of it, my understanding of the divide between the American left and the right moved from my head to my viscera.

We farm in the Sierra foothills sixty miles northeast of Sacramento, in the old mining town of Nevada City, California. We grow vegetables for the local market and a four-season diet including meat, dairy, grains, and beans for ourselves and a small community. If you were to visit Nevada City, you would see that it has the characteristics of a liberal hot spot: a thriving co-op, organic restaurants, a farmers' market, a thoroughly protected river, a community radio station of thirty-one years, and more yoga studios and little schools than you could shake a stick at. Even so, this is only a narrow part of the region's culture.

The nearby town of Rough and Ready seceded from the Union in 1850. It rejoined three months later, but the spirit that initiated the split has its modern manifestations. At the small, friendly, and professional butcher shop where we take our bacon and hams to get cured, a sign on the door boldly proclaims I'LL KEEP MY GUNS, FREEDOM, AND MONEY. YOU CAN KEEP THE "CHANGE." In step with the rest of inland California, Nevada County consistently elects Republican representatives. In 2008, we banded together, focused on the family, and by a margin of three votes "defended" marriage. The county lines were drawn in the shape of a Derringer. It's a region in which you can get your chakras aligned on the way to the shooting range. But no one does that, and that's the point.

Our district's town hall meeting fell on a hot and dry day in early September. We wanted to look like upstanding citizens, so the three of us cleaned up, trimmed our beards, and put on some decent clothes. We arrived twenty minutes early. The large venue was already approaching capacity and rang with animated exchanges. We found seats in a sea of red shirts and star-spangled accessories. I sat beside a woman who, after we had settled in a bit, looked me over and said in a tone so disdainful that I thought she had to be joking, "You look like a liberal."

I responded hesitantly and through a weak laugh, "Yes, I suppose I am." She let out a disgusted *harrumph* and moved to the next seat over.

I had to defend myself and my alleged liberalism from such immediate and damning disdain. My disbelief at her contempt quickly became anger and I soon found myself threatening to slap her. (She was the lunatic; she!) She proudly offered her cheek, disparaging liberals for always being victims. I groaned.

After five minutes of this — "Barack and all his czars. What more evidence do you need?" — our vehemence wore us out. I asked the woman if she kept a garden. She did, out of concern for Big Brother. Did I? Yes, because of my concerns for health and food security. Oh. We both let out a great exhale. This other person isn't a complete lunatic.

In agreeing upon the goodness of an action in pursuit of this shared value — autonomy, differently formulated — our many contrary conceptions about the world fell to the side. After a few more minutes of talking about gardening, I invited her back to her old seat and she accepted. Democracy in America was saved and I won a medal.

The meeting didn't amount to much. Congressman McClintock began the evening by taking a voice vote. The chorus in favor of the healthcare bill was loud and accented by clapping. That opposed was thunderous. Some heartfelt and thoughtful stories were shared, but generally, the various news-machine talking points were bandied back and forth until it was time to go home.

On the way out, as we waited to enter a slowly moving stream of people, a woman leaned toward us and enthusiastically commented, "Oh, nice costumes!" We looked to confirm if she was even talking to us (she was) and then looked at what we had thought were decent, public-suitable clothes: boots albeit shoddy ones, loose pants rolled up to the ankles because it had been hot, collared shirts of various accepted patterns and colors, and our short-brimmed caps. We looked back to her, confused; it was still very loud outside, and maybe we didn't hear her correctly.

She spoke again: "Aren't you guys dressed up? Like Bolsheviks, from that movie? You guys look dead-on like Bolsheviks!"

Before we had a chance to come up with a response, she was carried away by the flow of people, saying loudly, "Workers of the world unite!"

We left the meeting hall dumbstruck. What had just happened? We thought we looked nice, not like communists! Were the red-shirted Tea Party members not the only ones to have pushed so hard and ended up on the fringe?

The two experiences got me thinking. I don't want to be merely a farmer in a clearing of the woods; I want to be an active participant in the navigation of our community. If in our small, small-farm world, our notion of looking presentable summons a perceived alignment with leftist, revolutionary politics, aren't we unnecessarily obstructing the empathic, reasoned discussion necessary for a community to come together around common goals and pursue them to their conclusion? I don't want to exacerbate a sufficiently contentious environment by looking like a Bolshevik.

I also want to ask questions that are larger than those prompted by fashion, because it's obvious that our wardrobe malfunction was emblematic of the whole evening's disarray. And so, I ask: What aspects of Nevada County are these red-shirts proud of? Why did I not recognize them from the farmers' market? Would the culture of our open-to-all weekly potlucks really be welcoming to them no matter what they brought? What do they think of Community Supported Agriculture? Have they even heard of it? Do they listen to the longstanding community radio station? Do they take pride in the Yuba River? If the red-shirts didn't infer we were in costume, making a mockery of something, did they write us off for other reasons? What would it take to communicate that we're serious people, laboring through the best years of our lives to grow invigorating food that increases our community's autonomy and security and begins to make again possible a place-based local culture? Is it their responsibility to come to center or ours?

This project of finding common ground with people who voice conservative ideals would be a lot more daunting if our agrarianism wasn't an honest attempt to embody the most fundamental of conservative tenets: There are limits to everything. Given that, I believe local farming can be a rallying point for those on the left and those on right who refuse to believe that the other half of this country is made up of pea-brained aliens, despite the evidence so vehemently presented to the contrary. You know, by pundits.

So, in broad strokes, ecological farming is currently thought of as a liberal project. It's seen as a continuation of the seventies environmental movement in its consideration of the health of all beings in relation to human health. Those environmentalists pleaded for our society and economy to be realistic in their use of natural resources. This assertion of limits was critiqued by conservatives on the grounds that it was unrealistic in how it understood human needs. (A discussion about limits will always revolve around realism; each side is always saying, "Let's be realistic. . . .") So, whereas conservatives generally recognize social limits (such as the ability of the institution of marriage to accommodate an unfamiliar arrangement, for example) and economic limits (such as our ability to fund and manage a federal system of entitlements), they have deferred to liberals to fuss fecklessly about natural limits. When isolated and set against one another, it's clear which set of limits — social, economic, natural — guides the electorate.

A woman looked me over and said in a tone so disdainful that I thought she had to be joking, "You look like a liberal."

Ecological farming arrives on the scene, fulfilling the environmentalist objective of relating to the natural world within the understanding that there are limits that we surpass at our peril, and withstands the corresponding conservative critique by doing the real work within these limits to provide the human necessities for ourselves and our communities. (We aren't coming from the city and asking foresters to please, goddamnit, mind the pretty birds.)

All this is done while trending to the values of simplicity, commitment, and intimate, stable communities, all values associated more with the right side of the political spectrum. So, granted that bucking the dominant agricultural system takes a bit of the liberal world-can-be-better mentality, how many of our goals are liberal in the sense of being untested and perhaps foolishly optimistic?

We want to farm with methods and tools tested by hundreds of years; we want to do physical, skill-based work; and we want to actively and tangibly provide for a community, enabling an intimate connection to a place and its people. We gladly accept the challenge of creating rich lives within these constraints because we agree with the conservative sentiment that satisfaction and a life full of meaning cannot be achieved merely through professional accomplishment, the accumulation of toys, or even a long healthy life — great though all those are. Rather — as reluctant as I am to say something so pious — we believe that a life is made rich by one's relationships within a cohesive community. It's not food we're after; it's meals. To the liberal mentality that prioritizes the actualization of the self above all, we are saying that this has gone too far, and we are embracing family and community commitments. We are realistic in our belief that relationships within a community cannot be contingent upon mere chance or similarity, knowing from experience that a community without relationships based on economic ties of mutual assistance is loose and unsure of itself.

So, I think that even though we might dress like liberals, have been educated like liberals, create products generally bought by liberals, or come from liberal families or communities, at heart and in deed we are quite conservative. I do not want to attempt a takeover of conservative ideals, but rather to make small-scale, ecological agriculture available to self-identifying conservatives by recognizing that we too are tangibly pursuing these values — and we need their help. I don't think ecological agriculture is being used to its full potential for establishing new and potentially strong political alliances, not to mention cohesive communities. Once we recognize that we have the same goals, we can unify our means.

What recent generation has been so blessed with a clear purpose by which to focus its energies? To be given the task of blending the old and trusty (if a bit rusty) with the new and promising (if a bit plastic) to revivify the American promise of materially autonomous, self-governing communities with place-based traditions and cultures? I'm sure it was something spectacular to live in a time of old-growth forests and prairies. But what of the privilege to plant so many seeds and nurture as many saplings? Yes! Thank you! We want to work! We are ready to work. We will take them to adolescence. ✍

Resources

Erin Bullock
Mud Creek Farm
www.mudcreekfarm.com

Courtney Lowery Cowgill and Jacob Cowgill
Prairie Heritage Farm
www.prairieheritagefarm.com

Douglass DeCandia
Food Bank for Westchester
www.foodbankforwestchester.org

Evan Driscoll
Running with Pitchforks
www.runningwithpitchforks.com
Sasquatch Acre
http://sasquatchacre.com

Andrew French
Living the Dream Farm
http://ltdfarm.com

Adam Gaska
Mendocino Organics
www.mendoorganicscsa.com

Liz Graznak
Happy Hollow Farm
www.happyhollowfarm-mo.com

Brad Halm
Seattle Urban Farm Co.
www.seattleurbanfarmco.com

Lynda Hopkins
Foggy River Farm
www.foggyriverfarm.org
The Wisdom of the Radish
www.wisdomoftheradish.com

Sarah Hucka
Circle h Farm
www.circlehorganicfarm.com

Ben James
Town Farm
www.nohotownfarm.com

Neysa King
Dissertation to Dirt
www.dissertationtodirt.com

Samantha Lamb
Samantha Lamb Photography
www.samanthalamb.com

Sarahlee Lawrence
Plots to Plates
http://plotstoplates.wordpress.com
Rainshadow Organics
www.rainshadoworganics.com
River House: A Memoir
www.sarahleelawrence.com

Maud Powell
Siskiyou Sustainable Cooperative
www.siskiyoucoop.com

Meg Runyan
Wild Goose Farm
www.wildgoosefarm.net

Ginger Salkowski
Revolution Gardens
www.revolutiongardens.com

Sarah Smith
Grassland Farm
www.grasslandorganicfarm.com

Sarajane Snyder
Look at the Sky and Tell the Weather
http://fairweatherly.wordpress.com

A. M. Thomas
Wear a Wax Dustcoat
http://wearawaxdustcoat.com

Tanya Tolchin
On the Lettuce Edge
www.thelettuceedge.com

Josh Volk
joshvolk.com
www.joshvolk.com
Slow Hand Farm
www.slowhandfarm.com

Jenna Woginrich
Cold Antler Farm
http://coldantlerfarm.blogspot.com

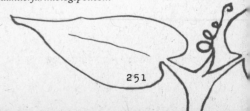

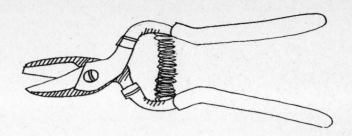

Recommended Reading

Katz, Sandor Ellix. *The Revolution Will Not Be Microwaved: Inside America's Underground Food Movements.* Chelsea Green Publishing, 2006.

Kimball, Kristin. *The Dirty Life: A Memoir of Farming, Food, and Love.* Scribner, 2010.

Kingsolver, Barbara. *Animal, Vegetable, Miracle: A Year of Food Life.* HarperCollins, 2007.

Pollan, Michael. *In Defense of Food: An Eater's Manifesto.* Penguin, 2008.

———. *The Omnivore's Dilemma: A Natural History of Four Meals.* Penguin, 2006.

Salatin, Joel. *The Sheer Ecstasy of Being a Lunatic Farmer.* Polyface, Inc. 2010.

Smith, Alisa and J. B. MacKinnon. *Plenty: One Man, One Woman, and a Raucous Year of Eating Locally.* Harmony Books, 2007.

Smith, Jeremy N. *Growing a Garden City: How Farmers, First Graders, Counselors, Troubled Teens, Foodies, a Homeless Shelter Chef, Single Mothers, and More Are Transforming Themselves and Their Neighborhoods Through the Intersection of Local Agriculture and Community — And How You Can, Too.* Skyhorse Publishing, 2010.

Farm Opportunities — Apprenticeships and Jobs

GrowFood
www.growfood.org
Runs an organic volunteer program that connects volunteers with farms

Northeast Beginning Farmers Project, Cornell University
http://nebeginningfarmers.org/farmers/learning-to-farm/farming-opportunities

NCAT Sustainable Agriculture Project
www.attra.ncat.org/attra-pub/internships

Backdoor Jobs
www.backdoorjobs.com/farming.html

Worldwide Opportunities on Organic Farms (WWOOF)
www.wwoof.org

Planning Worksheets

"Exploring the Small Farm Dream:
Is Starting an Agricultural Business
Right for You?"
New England Small Farm Institute
*www.smallfarm.org/main/for_new_
farmers/exploring_the_small_farm_dream*
  A decision-making workbook pub-
  lished by New England Small Farm
  Institute

"Self Assessment & Resource Assessment"
New England Small Farm Institute
*www.smallfarm.org/main/for_new_
farmers/resources_by_topic/self_and_
resource_assessment*

Worksheets
Northeast Beginning Farmers Project,
Cornell University
*http://nebeginningfarmers.org/farmers/
worksheets-2*

Business Planning

Born, Holly. "Agricultural Business
  Planning Templates and Resources,
  RL042" NCAT Sustainable
  Agriculture Project, June 2004.
*https://attra.ncat.org/attra-pub/
summaries/summary.php?pub=276*

Center for Farm Financial
  Management, University of
  Minnesota
www.agplan.umn.edu

Minnesota Institute for Sustainable
  Agriculture. "Building a Sustainable
  Business: A Guide to Developing a
  Business Plan for Farms and Rural

  Businesses." Sustainable Agriculture
  Research and Education, 2003.
*www.sare.org/publications/business/
business.pdf*

Tunnicliffe, Robin. "Business Planning
  for Small Scale Community Farming
  Enterprises." Community Farms
  Program, FarmFolk/CityFolk, 2009.
*http://farmfolkcityfolk.ca/programs/farm/
cf/business-plan.html*

Wiswall, Richard. *The Organic Farmer's
  Business Handbook: A Complete
  Guide to Managing Finances, Crops,
  and Staff — and Making a Profit.*
  Chelsea Green Publishing, 2009.

Land Access General Reference

The Greenhorns. "Land. Liberty. Sunshine. Stamina. A Mini Compendium of
  Resources for Beginning Farmers on the Topic of Finding Sustainable Land
  Tenure." Cornell University Cooperative Extension, 2010.
www.thegreenhorns.net/resources/GH_landtenureworkshop_minicompendium.pdf

American Farmland Trust
202-331-7300
www.farmland.org
  Non-profit land access help

Equity Trust, Inc.
413-863-9038
www.equitytrust.org

Farmland Information Center
800-370-4879
www.farmlandinfo.org

Land for Good
603-357-1600
www.landforgood.org

National Community Land Trust
Network
503-493-1000
www.cltnetwork.org

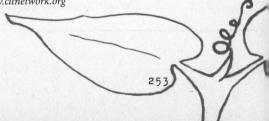

LAND-LINKING PROGRAMS

Network Participants
International Farm Transition Network
www.farmtransition.org/netwpart.html
  A comprehensive list of state-based land-linking programs

PROFITS AND PRICING

Born, Holly. "Enterprise Budgets and Production Costs for Organic Production,
  RL041." NCAT Sustainable Agriculture Project, 2004.
www.attra.ncat.org/attra-pub/summaries/summary.php?pub=187

Economics, Statistics, and Market Information System
United States Department of Agriculture
http://usda.mannlib.cornell.edu

Growing for Market
Fairplain Publications, Inc.
800-307-8949
www.growingformarket.com
  "News, advice and resources for market farmers"

Macher, Ron. *Making Your Small Farm Profitable*. Storey Publishing, 1999.

Organic Price Report
Rodale Institute
www.rodaleinstitute.org/organic-price-report

TAXES AND ACCOUNTING

Aubrey, Sarah Beth. *Starting & Running Your Own Small Farm Business. Storey Publishing, 2007.*

Beginning Farmer and Rancher Resources
http://beginingfarmerrancher.wordpress.com
  A resource blog by Poppy Davis

Salatin, Joel. *You Can Farm: The Entrepreneur's Guide to Start & Succeed in a Farming Enterprise.* Polyface, 1998.

Small Business/Self-Employed Virtual Small Business Tax Workshop
Internal Revenue Service, U.S. Department of the Treasury
www.tax.gov/virtualworkshop

USDA BEGINNING FARMER LOAN LITERACY

Farm Service Agency
United States Department of Agriculture
www.fsa.usda.gov
  The Farm Service Agency (FSA) provides direct and guaranteed loans to beginning farmers and ranchers who are unable to obtain financing from commercial credit sources. Contact them for information on their programs and to find your local FSA office.

National Council of State Agricultural Finance Programs
www.stateagfinance.org
  For information on state loan programs

Learning and Networking Conferences

EcoFarm Conference
Ecological Farming Association
831-763-2111
www.eco-farm.org
   Held annually in Pacific Grove,
   California

Farmer to Farmer Conference
Maine Organic Farmers and Gardeners
Association
www.mofga.org
   Annually in November; Northport,
   Maine

Farming for the Future Conference
Pennsylvania Association of Sustainable
Agriculture
www.pasafarming.org/our-work/farming-
for-the-future-conference
   Annually in February; State College,
   Pennsylvania

Georgia Organics Annual Conference
and Expo
www.georgiaorganics.org/conference.aspx
   Annually in March; various cities in
   Georgia

Healthy Farms & Rural Advantage
Conference
Nebraska Sustainable Agriculture Society
www.nebsusag.org
   Annually in February; various cities in
   Nebraska

MOSES Organic Farming Conference
Midwest Organic and Sustainable
Education Service
www.mosesorganic.org/conference.html
   Annually in February; La Crosse,
   Wisconsin

National Biodynamic Conference
Biodynamic Farming and Gardening
Association
888-516-7797
www.biodynamics.com
   Biannual; in various cities across the
   country

New Agrarian Conference
Quivira Coalition
www.quiviracoalition.org
   Annually in Albuquerque

NOFA Annual Winter Conference
Northeast Organic Farming Association
www.nofa.org
   Annually; in various Northeast states
   (Connecticut, Massachusetts, New
   Hampshire, New Jersey, New York,
   and Vermont)

NPSAS Winter Conference
Northern Plains Sustainable Agriculture
Society
www.npsas.org/events.html
   Annually in February; various cities in
   South Dakota

Organicology Conference
Organicology
www.organicology.org
   Annually in February; Portland,
   Oregon

PFI Annual Conference
Practical Farmers of Iowa
www.practicalfarmers.org
   Annually in January; various cities
   in Iowa

Practical Tools and Solutions for
Sustaining Family Farms Conference
Southern Sustainable Agriculture
Working Group
www.ssawg.org
   Annually in January

Texas Organic Farmers and Gardeners
Conference
Texas Organic Farmers and Gardeners
Association
www.tofga.org
   Annually in February; various cities
   in Texas

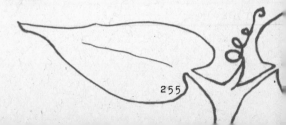

Recommended Reading about Sustainable Land Management

Butterfield, Jody, Sam Bingham, and Allan Savory. *Holistic Management Handbook: Healthy Land, Healthy Profits*. Island Press, 2006.

Dagget, Dan. *Gardeners of Eden: Rediscovering Our Importance to Nature*. Thatcher Charitable Trust, 2005.

Derosa, Lena Pellegrino. *Little Sixty: A Memoir of Growing up Italian-American in Appalachia*. Author House, 2010.

Dryland Solutions, Inc.
www.drylandsolutions.com

Imhoff, Daniel. *Farming with the Wild: Enhancing Biodiversity on Farms and Ranches*. Sierra Club Books, 2003.

Judy, Greg. *Comeback Farms: Rejuvenating Soils, Pastures and Profits with Livestock Grazing Management*. Green Park Press, 2008.

Nation, Allan. "Raw Organic Milk Appears to Revitalize Soil." Allan's Blog, *Stockman Grass Farmer*, 27. August 2009. *http://wincustomersusa.com/stockman*

Soil Doctor
www.soildoctor.org

Sponholtz, Craig and Doug Weatherbee. "Bringing Life Back to Your Land: Moisture, Microbes, and Climate Change." Quivira Coalition's 9th Annual Conference, "The Carbon Ranch: Using Food and Stewardship to Build Soil and Fight Climate Change." Albuquerque, 10 November 2010.

White, Courtney. "The Carbon Ranch: Fighting Climate Change One Acre at a Time." *Green Fire Times* (26 November 2010).

Whitten, George. "In the Mouth of the Tiger — Practicing Holistic Management on the Edge." *In Practice*, no. 102 (July/August 2005): 9–12.

Zeedyk, Bill and Van Clothier. *Let the Water Do the Work: Induced Meandering, an Evolving Method for Restoring Incised Channels*. Quivira Coalition, 2009.

Food Justice Organizations

America's Grow a Row
908-331-2962
www.americasgrowarow.org

American Community Gardening Association
877-275-2242
www.communitygarden.org

Community Food Security Coalition
503-954-2970
www.foodsecurity.org

FoodCorps, Inc.
info@foodcorps.org
http://foodcorps.org

Growing Power, Inc.
414-527-1546
www.growingpower.org

National Farm to School
www.farmtoschool.org

Slow Food USA
877-756-9366
www.slowfoodusa.org

Southern Sustainable Agriculture Working Group
479-251-8310
www.ssawg.org

Sustainable Food Center
512-236-0074
www.sustainablefoodcenter.org